# Algorithms for Efficient and Fast 3D-HEVC Depth Map Encoding

Gustavo Sanchez • Luciano Agostini
César Marcon

# Algorithms for Efficient and Fast 3D-HEVC Depth Map Encoding

Gustavo Sanchez
Centro de Informática
Instituto Federal Farroupilha (IFFAR)
Alegrete, Rio Grande do Sul
Brazil

Luciano Agostini
Video Technology Research Group (ViTech)
Federal University of Pelotas (UFPel)
Pelotas, Rio Grande do Sul
Brazil

César Marcon
Polytechnic School
Pontifical Catholic University of Rio
Grande do Sul (PUCRS)
Porto Alegre, Rio Grande do Sul
Brazil

ISBN 978-3-030-25929-7      ISBN 978-3-030-25927-3   (eBook)
https://doi.org/10.1007/978-3-030-25927-3

This Springer imprint is published by the registered company Springer Nature Switzerland AG
The registered company address is: Gewerbestrasse 11, 6330 Cham, Switzerland

# Preface

The rapid evolution of multimedia communications relying on services and applications that mostly use image and video in different formats has been driving worldwide research and engineering developments towards efficient coding of visual information. Three-dimensional (3D) visual information represented by color texture plus depth is one of such formats, which imposes quite demanding requirements on both hardware and software implementations of standard coding algorithms. Computational complexity, memory, bandwidth, and power consumption are constraining parameters with a great deal of impact on the performance of 3D video coding systems.

This book addresses this challenging field of research and engineering by presenting fast and efficient methods for encoding depth data of 3D visual information within the scope of the 3D extension of the high efficiency video coding (3D-HEVC). The book contains accurate technical descriptions of the main coding tools used for efficient compression of depth maps with particular emphasis on the prediction modes that are responsible for attaining significant compression performance. Jointly with the presentation of the test conditions that must be followed for benchmarking and validation of results, this is crucial and updated information about the standard that provides the necessary background for further analysis and technical developments. Several fast encoding algorithms are described throughout the book, departing from a deep analysis of the main contributing factors for the computational complexity and then proposing efficient solutions to speed up the coding process of depth information at negligible loss of efficiency.

The authors of the book have long research experience and significant contributions for the optimization of standard video coding systems. This highly specialized background also ensures that new methods described in the book are timely and useful for students and researchers interested in further investigating fast coding methods for 3D-HEVC algorithms and engineers developing efficient systems and applications, specially those including 3D multimedia in resource-constrained environments. Overall, this book presents valuable contributions to advance the technical design and implementation of fast and efficient 3D-HEVC standard encoders.

Pedro A. Amado de Assunção
Instituto de Telecomunicações
Politécnico de Leiria, Leiria, Portugal

# Acknowledgment

The authors thank their institutions in Brazil for supporting the development of this work, including the Pontifical Catholic University of Rio Grande do Sul (PUCRS), the Federal University of Pelotas (UFPel), and the Federal Institute of Education, Science, and Technology Farroupilha (IFFar), Alegrete Campus. The authors also thank the Brazilian research support agencies that financed this research: the Foundation for Research Support of Rio Grande do Sul (FAPERGS), the National Council for Scientific and Technological Development (CNPq), and the Coordination for the Improvement of Higher Education Personnel (CAPES).

# Contents

# Chapter 1
# Introduction

## 1.1 3D Video With Depth Map Coding

Several high-resolution video applications have arisen in the last decade demanding high efficiency and quality of encoding. Besides, these videos are stored in several media and places and streamed over several heterogeneous communication systems distributed at the internet. Therefore, video coding experts spent a high effort in the standardization of the modern Two-Dimensional (2D) video coding standards such as High Efficiency Video Coding (HEVC) [1], VP9 [2], Audio Video Coding Standard 2 (AVS2) [3], to obtain a high encoded video quality with a reduced stream size. However, currently, video coding utilization goes beyond capturing and encoding simple 2D scenes. Now, video applications allow sharing screens or enjoying a three-dimensional (3D) experience that goes beyond 2D videos by providing a depth perception of the scene. The 2D video coding standards do not encode these new video properties properly because they focus on the texture aspects of the scene; consequently, reducing the efficiency on capturing depth aspects of each video's scene. To fulfill this requirement, several HEVC extensions were designed by the video coding experts, including HEVC Screen Content (HEVC-SCC) [4], which enables achieving a higher performance when sharing computer screens or similar videos, and the 3D High Efficiency Video Coding (3D-HEVC) [5, 6], which better encode 3D video redundancies. Since this book focuses on 3D videos, we considered only 3D-HEVC in our explanations, which is the most efficient 3D video coding standard.

3D-HEVC was based in the Multiview Video plus Depth (MVD) data format [7], where each texture (conventional data seen in TV videos) frame is associated with a depth map captured from the same camera that captures the texture videos. The texture frames are the conventional images with colors that are presented at the TV, composed of luminance and chrominance samples. Gray shades using only luminance samples express the depth maps, which are the geometrical distance between

© Springer Nature Switzerland AG 2020
G. Sanchez et al., *Algorithms for Efficient and Fast 3D-HEVC Depth Map Encoding*, https://doi.org/10.1007/978-3-030-25927-3_1

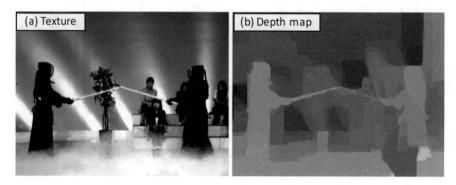

**Fig. 1.1** An example of (**a**) a texture view and (**b**) its associated depth map extracted from the Kendo video sequence [9]

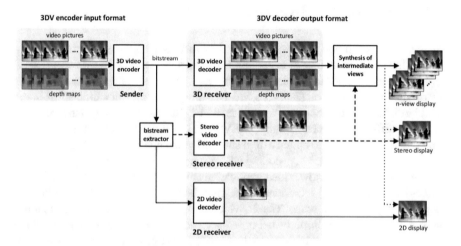

**Fig. 1.2** Example of 3D-HEVC encoding/decoding flow (based on [11])

the camera and the objects [8]. Figures 1.1a and 1.1b exemplifies a textured frame and its associated depth map, which were extracted from the Kendo video sequence [9].

The highest advantage of MVD is the stream size reduction for encoding a 3D video because the decoder can synthesize high-quality intermediary views interpolating texture views based on the depth data and using Depth Image Based Rendering (DIBR) [10] or others view synthesis techniques. These techniques allow synthesizing several high-quality intermediary texture views of the scene, reducing the number of stored/transmitted views. Figure 1.2 illustrates an abstraction of 3D-HEVC usage composed by coding and decoding processes. The first step is the scene capturing, then, the obtained data passes by the 3D-HEVC video encoding, followed by the video storing/transmitting. Next, the 3D, stereo and 2D video decoders decode the bitstream of the encoded video. Finally, according to the output video format, the intermediary virtual views are synthesized.

3D-HEVC requires multiple cameras to capture a scene, each one providing a raw texture and depth data captured from the same perspective. The entire data are encoded in a 3D-HEVC encoder, packing the data together inside an encoded stream. Each camera is called a view, where the texture and the depth content are included.

Several operations can be executed in the resulting encoded bitstream, which can be, for instance, stored in the device memory, sent for being stored in the cloud, broadcasted and/or transmitted. Whenever another application requires decoding this bitstream, the 3D-HEVC decoder reconstructs the original encoded views, providing the reconstructed raw texture and depth data. However, several 3D video applications demand much more than the small number of transmitted views; therefore, innumerous texture views located between the original encoded views are synthesized using view-synthesis techniques. Notice, a stereo or 2D video decoder can also decode the 3D-HEVC bitstream without using the encoded depth maps.

Figure 1.1 illustrates depth maps contains large regions with homogeneous values (in both background and objects bodies) and sharp edges (in the borders of the objects), which is the main difference between the depth maps and texture characteristics. Considering these characteristics, the video coding experts designed algorithms that can encode depth maps with higher efficiency and quality. For obtaining a high-quality synthesized-view, a quality of the encoded depth maps is essential, which leads to a significant challenge in this scenario since encoding depth maps using only traditional 2D video coding techniques (without considering their edges) introduces encoding errors.

The viewers do not visualize the quality of the depth map encoding; however, it affects the 3D video quality indirectly since depth maps are used to synthesize virtual views that are watched by the application viewers. The highest difference in the synthesized views quality is noticed in the depth map edges; therefore, it is essential to encode the depth maps as well as possible, generating depth map edges without errors in the video synthesis process [12].

Considering the importance of edge preservation and the desire of obtaining high compression rates while maintaining high video quality, the Joint Collaborative Team on 3D Video Coding Extension Development (JCT-3 V) experts have designed new tools for 3D-HEVC depth map encoding. In intra-frame prediction, the new encoding tools are the Intra-picture skip [13], Direct Component-only (DC-only) [14] and the bipartition modes (composed of the Intra_Wedge and Intra_Contour [15] algorithms). Besides, it is essential to emphasize that during the 3D-HEVC standardization, the names of these tools have changed; therefore, solutions found in the literature may be refereeing to the old algorithm. The bipartition modes are also described in the literature as Depth Modeling Modes (DMMs), where Intra_Wedge and Intra_Contour are designed as DMM-1 and DMM-4, respectively. The Intra-picture skip and DC-only are also described in the literature as Depth Intra Skip (DIS) and Segment-wise DC Coding (SDC), respectively.

Considering that 3D-HEVC was the first video standard to use MVD, depth map coding has not been as extensively explored in the literature as the coding of textures over the last decade, having significant space for researchers to provide new contributions in this area. Besides, the simplicity of depth maps was not considered when several encoding tools were inherited from the HEVC texture coding, and most of the new tools that were designed, especially for depth maps encoding, were not the target of significant research exploration; therefore, opening space for more in-depth exploration of the depth map coding. Thus, this Book describes our exploration in depth map characteristics and coding algorithms in several levels aiming to increase the efficiency of the depth map coding implementations.

## 1.2   Book Organization and Contributions

This book is composed of six chapters that present the concepts and essential contributions to the 3D-HEVC coding. The second chapter provides background information about 3D-HEVC depth map encoding. Initially, the 3D-HEVC encoding structure is described in detail, relating its frame and quadtree structures. The algorithms used in the intra-frame prediction are described in detail so that the reader can understand the significant differences in the intra-frame prediction of texture and depth maps. Then, inter-frame and inter-view prediction algorithms are described, followed by the description of standard tools between intra- and inter-frame/view predictions. Chapter 2 ends by describing the Common Test Conditions (CTC), which contains the tests that researchers and developers must follow to generate results that can be compared to related work.

Chapter 3 presents an overview of the state-of-the-art research on computational effort reduction algorithms, including intra-frame, inter-frame and inter-view predictions.

Chapter 4 shows an in-depth analysis of the encoder processing effort and the usage of the encoding tools; based on them, four new timesaving encoding algorithms are presented. The new algorithms include two solutions for accelerating the Intra_Wedge, one solution for accelerating the Transform-Quantization (TQ) and DC-only encoding flows, and one solution for pruning the quadtree structure without requiring an extensive evaluation. The experimental results are presented, being capable of providing several levels of timesaving with different levels of impact in the encoding efficiency. The contributions of this chapter can be summarized as follows:

- Detailed analysis of the encoding time distribution in the encoding tools for depth map intra-frame prediction [16, 17];
- Detailed analysis of the encoding mode usage in the depth map intra-frame prediction [17];
- Two solutions for accelerating the Intra_Wedge mode [18, 19];
- One solution for accelerating TQ and DC-only encoding flows [20];
- One solution for pruning the quadtree without requiring an extensive evaluation [21].

Chapter 5 presents the algorithms developed for encoding time reduction of the depth map inter-frame prediction. Three algorithms are presented, along with their motivational analysis, where each algorithm actuate in a different encoding level. The first algorithm considers depth map simplicity, targeting both Motion Estimation (ME) and Disparity Estimation (DE). The second proposal focuses on reusing the texture encoding information for selecting a lower quantity of tools for being evaluated if these tools can obtaining sound Rate-Distortion results. The last algorithm uses decision trees built employing machine learning for pruning the quadtree evaluation when lower levels of quadtree are not required. The contributions of this chapter can be summarized as follows:

- One solution for accelerating ME/DE in the depth map coding [22];
- One solution using inter-component data for accelerating the depth map coding [23];
- One solution with machine learning for pruning the quadtree evaluation [24].

Chapter 6 summarizes the conclusions and the contributions described along with this Book. The importance of reducing the encoding effort of the 3D-HEVC depth maps is highlighted, and a brief description of the main contributions are also presented. Besides, we present and discuss open research possibilities.

# References

1. Sullivan, G.J., J.-R. Ohm, W.-J. Han, T. Wiegand, et al. 2012. Overview of the high efficiency video coding (HEVC) standard. In *IEEE transactions on circuits and systems for video technology*, vol. 22–12, 1649–1668.
2. He, Z., L. Yu, X. Zheng, S. Ma, and Y. He. 2013. Framework of AVS2-video coding. In *IEEE International Conference on Image Processing*, 1515–1519.
3. Mukherjee, D., J. Bankoski, A. Grange, J. Han, J. Koleszar, P. Wilkins, Y. Xu, and R. Bultje. 2013. The latest open-source video codec VP9-an overview and preliminary results. In *Picture coding symposium*, 390–393.
4. Xu, J., R. Joshi, and R.A. Cohen. 2016. Overview of the emerging HEVC screen content coding extension. *IEEE Transactions on Circuits and Systems for Video Technology* 26 (1): 50–62.
5. Müller, K., H. Schwarz, D. Marpe, C. Bartnik, S. Bosse, H. Brust, T. Hinz, H. Lakshman, P. Merkle, F.H. Rhee, et al. 2013. 3D high-efficiency video coding for multi-view video and depth data. *IEEE Transactions on Image Processing* 22 (9): 3366–3378.
6. Tech, G., Y. Chen, K. Müller, J.-R. Ohm, A. Vetro, and Y.-K. Wang. 2016. Overview of the multiview and 3D extensions of high efficiency video coding. *IEEE Transactions on Circuits and Systems for Video Technology* 26 (1): 35–49.
7. Kauff, P., N. Atzpadin, C. Fehn, M. Müller, O. Schreer, A. Smolic, and R. Tanger. 2007. Depth map creation and image-based rendering for advanced 3DTV services providing interoperability and scalability. *Signal Processing: Image Communication* 22 (2): 217–234.
8. Zhao, Y., C. Zhu, Z. Chen, D. Tian, and L. Yu. 2011. Boundary artifact reduction in view synthesis of 3D video: From perspective of texture-depth alignment. *IEEE Transactions on Broadcasting* 57 (2): 510–522.
9. Tang, X.-l., S.-K. Dai, and C.-H. Cai. 2010. An analysis of TZ Search algorithm in JMVC. In *International Conference on Green Circuits and Systems*, 516–520.

10. Fehn, C. 2004. Depth-image-based rendering (DIBR), compression, and transmission for a new approach on 3D-TV. In *Stereoscopic displays and virtual reality systems XI*, 93–105.
11. Chen, Y., G. Tech, K. Wegner, and S. Yea. 2015. Test model 11 of 3D-HEVC and MV-HEVC, Technical Report, ISO/IEC JTC1/SC29/WG11, 58p.
12. Smolic, A., K. Muller, K. Dix, P. Merkle, P. Kauff, and T. Wiegand. 2008. Intermediate view interpolation based on multiview video plus depth for advanced 3D video systems. In *IEEE International Conference on Image Processing*, 2448–2451.
13. Lee, J., M. Park, and C. Kim 2015. 3d-ce1: depth intra skip (dis) mode, Technical Report, ISO/IEC JTC1/SC29/WG11, 5p.
14. Liu, H., and Y. Chen. 2014. Generic segment-wise DC for 3D-HEVC depth intra coding. In *IEEE International Conference on Image Processing*, 3219–3222.
15. Merkle, P., K. Müller, D. Marpe, and T. Wiegand. 2016. Depth intra coding for 3D video based on geometric primitives. *IEEE Transactions on Circuits and Systems for Video Technology* 26 (3): 570–582.
16. Sanchez, G., R. Cataldo, R. Fernandes, L. Agostini, and C. Marcon. 2016. 3D-HEVC depth prediction complexity analysis. In *IEEE International Conference on Electronics, Circuits, and Systems*, 348–351.
17. Sanchez, G., J. Silveira, L. Agostini, and C. Marcon. 2018. Performance analysis of depth intra coding in 3D-HEVC. *IEEE Transactions on Circuits and Systems for Video Technology* 1 (1): 1–12.
18. Sanchez, G., L. Agostini, and C. Marcon. 2018. A reduced computational effort mode-level scheme for 3D-HEVC depth maps intra-frame prediction. *Journal of Visual Communication and Image Representation* 54 (1): 193–203.
19. Sanchez, G., L. Jordani, C. Marcon, and L. Agostini. 2016. DFPS: A fast pattern selector for depth modeling mode 1 in three-dimensional high-efficiency video coding standard. *Journal of Electronic Imaging* 25 (6): 063011.
20. Sanchez, G., L. Agostini, and C. Marcon. 2017. Complexity reduction by modes reduction in RD-list for intra-frame prediction in 3D-HEVC depth maps. In *IEEE International Symposium on Circuits and Systems*, 1–4.
21. Saldanha, M., G. Sanchez, C. Marcon, and L. Agostini. 2018. Fast 3D-Hevc depth maps intra-frame prediction using data mining. In *IEEE International Conference on Acoustics, Speech and Signal Processing*, 1738–1742.
22. Sanchez, G., M. Saldanha, B. Zatt, M. Porto, L. Agostini, and C. Marcon. 2017. Edge-aware depth motion estimation—A complexity reduction scheme for 3D-HEVC. In *European Signal Processing Conference*, 1524–1528.
23. Saldanha, M., G. Sanchez, C. Marcon, and L. Agostini. 2018. Block-level fast coding scheme for depth maps in three-dimensional high efficiency video coding. *Journal of Electronic Imaging* 27 (1): 010502.
24. ———. 2019. Fast 3D-HEVC depth maps encoding using machine learning. *IEEE Transactions on Circuits and Systems for Video Technology* 1–1: 12.

# Chapter 2
# 3D-HEVC Background

This chapter is divided into five sections containing the theoretical background of 3D-HEVC depth map encoding employed on this Book. Section 2.1 describes the 3D-HEVC basic encoding structure. Section 2.2 presents the intra-frame prediction algorithms used in the depth map encoding. The inter-frame and inter-view encoding algorithms are detailed in Sect. 2.3. Section 2.4 discusses the necessary encoding steps applied in the data, which is produced by the prediction algorithms. Finally, Sect. 2.5 describes the evaluation methodology and the CTC employed in 3D-HEVC evaluations.

## 2.1 3D-HEVC Basic Encoding Structure

The 3D-HEVC encoding is based on a smart prediction structure and a flexible quadtree structure. The prediction structure defines types of frames, which employ specific predictions strategies. The quadtree structure allows evaluating different block sizes, searching for the best combination of block partitioning.

### 2.1.1 3D-HEVC Temporal and Spatial Prediction Structure

Figure 2.1 shows the 3D-HEVC temporal and spatial prediction structure. The access unit is a basic element used in the prediction structure, where all textures and their associated depth maps are encoded for each k instant of time ($Time_k$).

The *dependent texture views* ($T_1$, $T_2$) of each access unit use the *base texture view* ($T_0$) as the reference for encoding the redundancies among views. $T_0$ is encoded directly by an HEVC encoder without any dependency with other views since an HEVC decoder can provide images for the 2D display from a 3D-HEVC bitstream. The dependent views are encoded using data from $T_0$ to explore the inter-view

© Springer Nature Switzerland AG 2020
G. Sanchez et al., *Algorithms for Efficient and Fast 3D-HEVC Depth Map Encoding*, https://doi.org/10.1007/978-3-030-25927-3_2

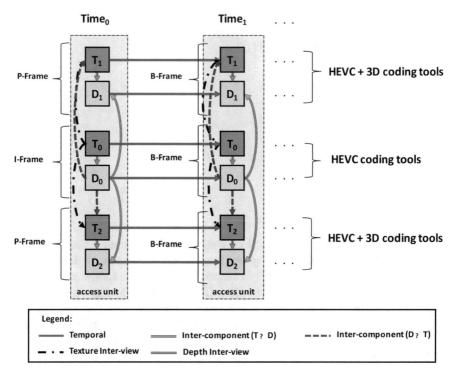

**Fig. 2.1** Basic prediction structure of 3D-HEVC [1]

redundancy. Figure 2.1 also shows the 3D-HEVC inter-frame predictions, allowing exploring inter-frame redundancies among access units; i.e., exploring the redundancies among the views captured at different times.

3D-HEVC defines three types of frames: I-, P-, and B-frames. I-frames are applied only in the base view ($T_0$), and they are encoded using the data inside the current encoding frame, i.e., employing the intra-frame prediction algorithms. I-frames are used to encode few base views – in the first video frame and from time to time (a specific amount of time defined by the encoder) to refresh the current scene. Consequently, I-frames allows reconstructing the video without any information of the past-encoded frames; therefore, I-frames can recover the video quality that eventually was lost among several frames propagation. Besides, I-frames can also allow the decoder to recover lost packages from transmission faults. P-frames are employed in the same access unities that encoded $T_0$ using I-frames. They can be encoded applying intra-frame and inter-view predictions to explore intra-frame and inter-view redundancies. Finally, B-frames can be encoded using intra-, inter-frame, and inter-view predictions. Inter-layer prediction techniques can also be applied in all types of frames.

The depth map ($D_0$, $D_1$, or $D_2$) has the same frame type of its associated texture frame; however, 3D-HEVC allows using other encoding tools such as Intra_Wedge, Intra_Contour, Direct Component-only (DC-only), and Intra-picture skip.

### 2.1.2   Quadtree Structure of 3D-HEVC

The 3D-HEVC depth map coding inherited the same quadtree structure of the HEVC texture coding [2]; i.e., before being encoded, the current frame is divided into Coding Tree Units (CTUs) – a smaller and basic coding structure. Subsequently, each CTU is split into four Coding Units (CUs), which can be divided recursively into four CUs until reaching a minimum pre-determined quadtree level.

Figure 2.2 illustrates a 64 × 64 CTU partitioning, where the encoder selects the final quadtree structure. It is essential to notice that this flexibility allows achieving a high encoding efficiency in several video coding standards since regions with fewer details can be encoded with larger blocks, while regions with more details can be further refined applying smaller blocks.

A CU can also be divided into Prediction Units (PUs), where intra-, inter-frame, and/or inter-view predictions encode the blocks considering the type of the encoding frame. The intra-frame prediction encodes a PU investigating the information contained inside the encoding frame. The inter-frame prediction explores the temporal redundancy; therefore, this tool uses past encoded frames in the same view as the reference to make inter-frame predictions. The inter-view and inter-frame predictions are similar; however, the inter-view prediction uses the redundant information in other views inside the access unit as a reference, instead of past encoded frames.

Figure 2.3 shows the possibilities of dividing the PU into several sizes according to the encoding type. Notice the PUs can assume only quadratic sizes (i.e., blocks with the same width and height) [3] in the intra-frame prediction. While in the inter-frame and inter-view predictions, rectangular partitions are also allowed.

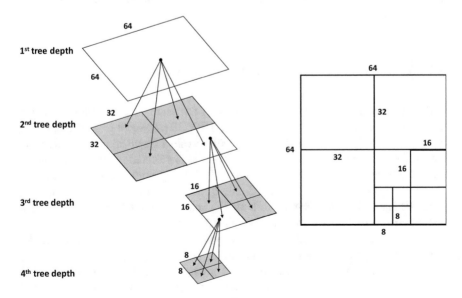

**Fig. 2.2**  3D-HEVC quadtree structure

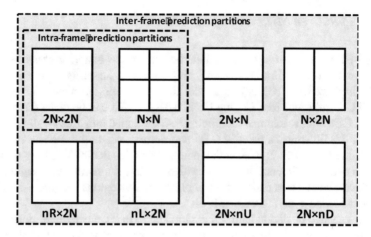

**Fig. 2.3** Partitioning of PUs

In these three predictions, the maximum and minimum allowed block sizes are 64 × 64 and 4 × 4, respectively.

The Rate Distortion-cost (RD-cost) is a function that combines the visual quality and the number of bits needed for a given tool to encode a block. The encoder computes the RD-cost of each block size, aiming to select the best available tool for encoding the block. By combining the RD-cost of smaller blocks, the encoder can evaluate several partitions possibilities and find the partition that leads to the smaller RD-cost for the resulting quadtree, i.e., the best possibility of encoding for that block. A high computational effort is required since the quadtree evaluates several encoding modes and block sizes.

## 2.2   Intra-Frame Prediction of Depth Maps

Figure 2.4 displays the intra-frame prediction encoding of the 3D-HEVC depth maps for a given depth block used in the 3D-HEVC Test Model (3D-HTM), which is the 3D-HEVC reference software [4]. After evaluating all modes according to their RD-cost, the mode with the lowest RD-cost is selected to encode the block. Each encoded block is assessed using: (i) HEVC intra-frame prediction in Transform-Quantization (TQ) and DC-only flows; (ii) bipartition modes (grouping Intra_Wedge and Intra_Contour) in both TQ and DC-only flows, and; (iii) Intra-picture skip mode directly applying Entropy Coding (EC).

The 3D-HEVC intra prediction tool was inherited from the texture intra prediction in HEVC. Both, Intra_Wedge and Intra_Contour [5] segment the encoding block into two regions, and each region is predicted with the Constant Partition Value (CPV) applying the average of all values of the region. Intra_Wedge employs a straight line called wedgelet to divide the encoding block, while the Intra_Contour tool creates a contour segmentation dynamically. The bipartition modes creation was focusing on achieving high efficiency when encoding edge regions.

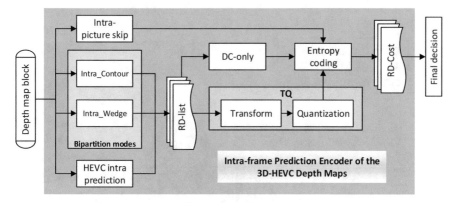

**Fig. 2.4** Intra-frame prediction model of the 3D-HEVC depth maps

The encoding tool used in the Intra-picture skip stage applies four new prediction modes for obtaining high bitrate reduction at homogeneous regions [6].

The HEVC intra-frame prediction and the bipartition modes utilize a Rate Distortion-list (RD-list) to determine some prediction modes that are further evaluated by their RD-cost since for assessing all encoding possibilities by their RD-cost demands a prohibitive computation effort. After processing the prediction tools, all modes included in the RD-list are evaluated by the TQ and DC-only flows. Subsequently, the RD-costs are achieved using entropy coding in the results of TQ and DC-only flows. The TQ flow was inherited from the HEVC texture coding, and the DC-only flow [7] has been designed as a new encoding alternative to the TQ flow, focused on achieving higher efficiency in the homogeneous regions of the depth maps. Additionally, the entropy encoder is applied in all Intra-picture skip modes to achieve the RD-cost without passing through the RD-list evaluation.

Following subsections detail the intra-frame prediction tools used on the depth map encoding. Section 2.4 describes the standard encoding tools between intra-, inter-frames and inter-views predictions (i.e., TQ and DC-only flow, and Entropy Coding).

## 2.2.1   HEVC Intra-Frame Prediction

The intra-frame prediction of the 3D-HEVC depth maps determines the planar, DC and, 33 directional modes, whose directions are presented in Fig. 2.5. Samples of spatially neighboring blocks are used as references in these modes for creating a predicted block [8].

A different number of directions is applied according to the block sizes when using HEVC intra-frame prediction; i.e., (i) 18 directions for 4 × 4 blocks, (ii) 35 directions for 8 × 8, 16 × 16 and 32 × 32 blocks; and (iii) 4 directions for 64 × 64 blocks.

Assessing the RD-cost of all available prediction modes is prohibitive for real-time applications. Therefore, 3D-HTM employs the heuristic introduced by Zhao et al. [9] to encode texture and depth maps. An RD-list is generated (see Fig. 2.4)

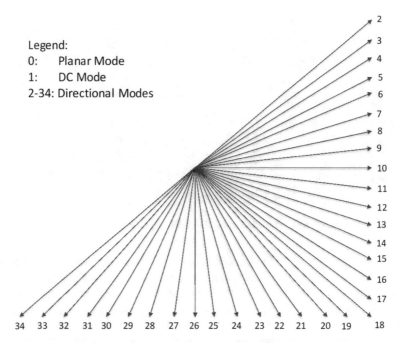

**Fig. 2.5** Prediction modes of the HEVC intra-frame (adapted from [8])

containing only a few modes, and the RD-cost is calculated only for these chosen modes. Two algorithms are used in this heuristic: Rough Mode Decision (RMD) and Most Probable Modes (MPM). The RMD algorithm applies the Sum of Absolute Transformed Differences (SATD) between the original block and the predicted one to evaluate all HEVC modes (without the entire RD-cost calculation) fastly. The algorithm sorts the intra-frame prediction modes according to their SATDs and injects the modes with the lowest SATDs into the RD-list (8 modes for 4 × 4 and 8 × 8 blocks, and 3 modes for 16 × 16, 32 × 32, and 64 × 64 blocks). Subsequently, the MPM algorithm checks the information of the modes used in the left and above encoded neighbor blocks and inserts up to two modes into the RD-list.

### 2.2.2  The Bipartition Modes

The bipartition modes use the Intra_Wedge and Intra_Contour algorithms for block sizes from 4 × 4 to 32 × 32, which produce the wedgelet segmentation (Intra_Wedge) exemplified in Fig. 2.6, and the contour segmentation (Intra_Contour) presented in Fig. 2.7. Notice these modes are not available for 64 × 64 blocks.

Figure 2.6a shows that Intra_Wedge divides the encoding block into two regions separated by a wedgelet, while Fig. 2.6b illustrates the discretization of Fig. 2.6a to select a region of these samples.

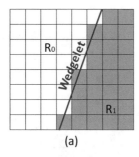

| 0 | 0 | 0 | 0 | 0 | 0 | 1 | 1 |
| 0 | 0 | 0 | 0 | 0 | 1 | 1 | 1 |
| 0 | 0 | 0 | 0 | 0 | 1 | 1 | 1 |
| 0 | 0 | 0 | 0 | 0 | 1 | 1 | 1 |
| 0 | 0 | 0 | 0 | 1 | 1 | 1 | 1 |
| 0 | 0 | 0 | 0 | 1 | 1 | 1 | 1 |
| 0 | 0 | 0 | 0 | 1 | 1 | 1 | 1 |
| 0 | 0 | 0 | 1 | 1 | 1 | 1 | 1 |

(a)                                                                    (b)

**Fig. 2.6** Wedgelet segmentation model of a depth map block; (**a**) Pattern with the regions $R_0$ and $R_1$, and (**b**) discretization of the regions with constant values

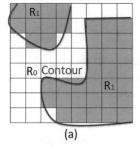

| 1 | 1 | 1 | 1 | 0 | 0 | 0 | 0 |
| 1 | 1 | 1 | 1 | 0 | 1 | 1 | 1 |
| 0 | 1 | 1 | 0 | 0 | 1 | 1 | 1 |
| 0 | 0 | 0 | 0 | 0 | 1 | 1 | 1 |
| 0 | 0 | 0 | 0 | 0 | 1 | 1 | 1 |
| 0 | 0 | 1 | 1 | 1 | 1 | 1 | 1 |
| 0 | 0 | 1 | 1 | 1 | 1 | 1 | 1 |
| 0 | 0 | 0 | 1 | 1 | 1 | 1 | 1 |

(a)                                                                    (b)

**Fig. 2.7** Contour segmentation model of a depth map block; (**a**) Pattern with the regions $R_0$ and $R_1$ and (**b**) discretization of the regions with constant values

**Table 2.1** Storage requirements and the number of wedgelets evaluated in the Intra_Wedge tool

| Block size | Number of available wedgelets | Wedgelets in the initial evaluation set | Storage requirements (bits) |
| --- | --- | --- | --- |
| $4 \times 4$ | 86 | 58 | 1,376 |
| $8 \times 8$ | 802 | 314 | 51,328 |
| >= $16 \times 16$ | 510 | 384 | 130,560 |

The Intra_Wedge algorithm determines the evaluation of several wedgelets for each block; however, only a subset is assessed, aiming to decrease the Intra_Wedge computational effort. Table 2.1 shows the possible wedgelets and the storage requirement (in bits) according to the block size.

Figure 2.8 exhibits a high-level diagram of the Intra_Wedge encoding algorithm applied in the 3D-HTM. Main, Refinement and Residue are the stages that compose the Intra_Wedge algorithm. The Main stage assesses the initial wedgelet set to find the best wedgelet partition among the available ones, which requires mapping the encoding block into the binary pattern defined by each wedgelet. According to this mapping, the average values of all samples of the depth maps mapped into regions R0 and R1 are computed, and a predicted block is defined as the average value of each region in the Prediction step. Next, the encoding

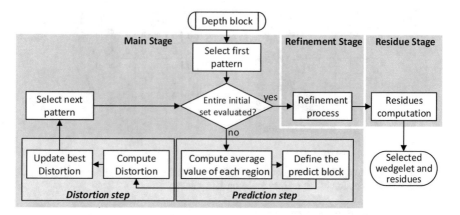

**Fig. 2.8** Main blocks of the Intra_Wedge encoding algorithm

algorithm applies a similarity criterion to determine the encoding efficiency of each pattern, creating a local distortion value. All distortions are compared, and the pattern with the lowest distortion is chosen as the best wedgelet in the Distortion step. The Synthesized View Distortion Change (SVDC) is used in the 3D-HTM as default Distortion metric [10].

The Intra_Wedge Refinement Stage evaluates at most eight wedgelets with a similar pattern than the wedgelet selected in the Main Stage. Once more, the best wedgelet is the one with the lowest distortion. Finally, the Residue Stage subtracts the predicted block of the elected wedgelet from the original one and appends this wedgelet into the RD-list.

The Intra_Contour is a contour segmentation that can model arbitrary patterns in two regions and even consist of various parts. Figure 2.7a exemplifies the contour segmenting a block, while Fig. 2.7b displays the contour discretization with the constant value of each region. Intra_Contour employes the inter-component prediction technique to find the best contour partition, using the previously encoded information from the texture during the prediction of depth maps [11].

JCT-3V experts designed the Intra_Contour algorithm driven by the fact that the depth block represents the same scene at the same viewpoint of the texture block, which is previously encoded. Although depth maps and texture contain distinct characteristics, they have high structural similarities that the encoding process can explore. For instance, an edge in the depth component usually corresponds to an edge in the texture component. Figure 2.9 highlights these similarities and correlations.

Figure 2.10 depicts the Intra_Contour encoding flow, which is composed of three stages: Texture Average, Prediction, and Residue. At the beginning of the Intra_Contour execution, the encoder only knows the reconstructed texture block and the current block of the encoding depth map. The Texture Average stage starts computing the average value (u) of the corners samples in the texture block.

The pseudo-code described in Fig. 2.11 is applied to construct a binary map that determines the Intra_Contour partition. Next, the algorithm maps all texture samples smaller than u into $R_0$, while the others are mapped into $R_1$, generating a binary map.

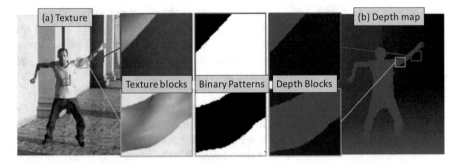

**Fig. 2.9**   Correlations and similarities between texture and depth maps (adapted from [25])

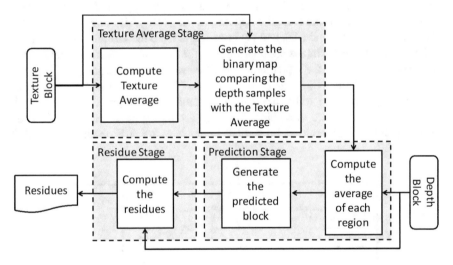

**Fig. 2.10**   The Intra_Contour encoding flow composed of three stages

**Fig. 2.11**  Pseudo-code describing the mapping in the Intra_Contour algorithm (blocks of N × N samples)

```
1.   for (i ← 1 to N)
2.       for (j ← 1 to N)
3.           if (sample[i][j] < u)
4.               R₀ ← sample[i][j]
5.           else
6.               R₁ ← sample[i][j]
```

The Intra_Contour encoding does not need to use depth information to create the binary map. Besides, the Intra_Contour pattern is not transmitted because the decoder can use the texture block, which is decoded before the depth block, and adaptively produce the Intra_Contour pattern at the decoder side, consequently reducing the bitrate. The Prediction stage processes the depth maps using this binary map and calculates the average values of all samples of the depth block in the regions $R_0$ and $R_1$. These average values are selected as the predicted block, according to the binary map. Finally, the residue is generated by subtracting the predicted block from the original depth block and inserting this mode into the RD-list, in the Residue stage.

The bipartition modes can produce CPVs that require several bits for transmission and storage. Thus, the 3D-HEVC standard adopted the delta CPV encoding, which computes the predicted CPV using the information of the neighbor pixels to eliminate this problem. Figure 2.12 detaches in blue the encoding block, showing the samples used in delta CPV computation.

The CPV prediction requires the (i) binary pattern information of the three corners positions (P0, P1, and P2), (ii) three samples of the upper PU block (TL, TM, and TR), (iii) one sample from the upper right PU (TRR); (iv) three samples from the left PU (LT, LM, and LB), and (v) one sample from the left bottom PU (LBB). When some of these samples are missing (i.e., the necessary neighbor pixel has not been codified yet), the CPV prediction uses the nearest available sample.

Figure 2.13 depicts the pseudo-code for delta CPV creation that assigns the references ref_0 and ref_1 to the matching CPV prediction (i.e., pred_0 and pred_1). Case 1 occurs when the three binary patterns are equal; then, ref_0 is calculated applying the average value of TL and LT. In this case, ref_1 is obtained from LBB

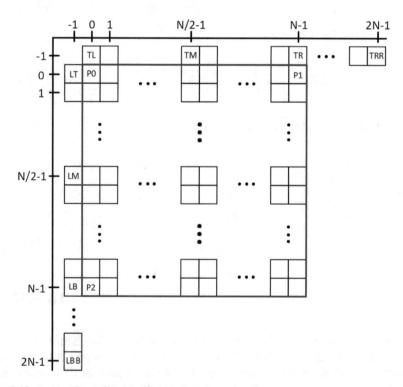

**Fig. 2.12** An N × N encoding block and its boundary samples

1.  if (P0 = P1 and P0 = P2)  $\longrightarrow$ Cases 1a, 1b
2.      ref_0 $\leftarrow$ average(TL, LT)
3.      if (abs(TRR − TL) > abs(LBB − LB))
4.          ref_1 $\leftarrow$ TRR $\longrightarrow$ Case 1a
5.      else
6.          ref_1 $\leftarrow$ LBB  $\longrightarrow$ Case 1b
7.  elsif (P0 ≠ P1 and P1 = P2)  $\longrightarrow$ Case 2
8.      ref_0 $\leftarrow$ average(TL, LT)
9.      ref_1 $\leftarrow$ average(LB, TR)
10. else                                    $\longrightarrow$ Case 3
11.     ref_0 $\leftarrow$ (P0 ≠ P2) ? TM : LM
12.     ref_1 $\leftarrow$ (P0 ≠ P2) ? LB : LR    $\longrightarrow$ Case 4
13. pred_0 $\leftarrow$ (P0 ≠ P2) ? ref_1 : ref_0
14. pred_1 $\leftarrow$ (P0 ≠ P2) ? ref_0 : ref_1

**Fig. 2.13** Pseudo-code for delta CPV computation

or TRR, according to the highest absolute difference between TRR and TL or LBB and LB. When P0 is different of P1 and P2 Cases 2 is chosen; this case uses the average value of the pairs (TL, LT) and (TR, LB) to compute ref_0 and ref_1, respectively. Case 3 happens when P0 is equal to P1 but different from the pattern of P2. In this case, TM is selected as ref_0 and LB as ref_1. Finally, Case 4 occurs when P0 is equal to P2 but differs from the pattern of P1; then, LM is selected as ref_0 and TR as ref_1.

The delta CPVs are computed by subtracting the original CPV from the predicted CPV, and only delta CPVs are transmitted in the bitstream. The predicted CPV can be calculated, when decoding the block, using the same algorithm used in the encoding process. It needs accessing the wedgelet memory to get the P0, P1, and P2 values, or producing the Depth Modeling Mode-4 (DMM-4) pattern to request the same sample values of neighbor samples employed by the encoder. The predicted CPV is attached with the delta CPV, reconstructing the original CPV.

### 2.2.3  Intra-Picture Skip

The Intra-picture skip was developed considering that most of the depth maps are smooth areas for block sizes ranging from 8 × 8 to 64 × 64. In general, slight modifications of depth map values in smooth areas do not affect the quality of the synthesized views significantly [6]. Thus, the Intra-picture skip ignores the residual coding for the smooth areas rendering a meaningful bitrate reduction.

The Intra-picture skip includes four prediction modes [6]: (i) Mode 0 (vertical intra-frame); (ii) Mode 1 (horizontal intra-frame); (iii) Mode 2 (single vertical depth); and (iv) Mode 3 (single horizontal depth). Mode 0 and Mode 1 copy all neighbor upper and left samples, respectively, to the current encoding block; whereas, Mode 2 and Mode 3 copy only the neighbor middle upper and left samples, respectively, to the current encoding block.

Figure 2.14a, b, c, d exemplify the Intra-picture skip Modes 0, 1, 2 and 3, respectively, concerning an 8 × 8 encoding block; red boxes represent the predicted blocks. Considering N2 the number of samples of the square encoding block, the prediction uses only the N above and N left neighbor samples. The HEVC vertical (Fig. 2.5, direction 26) and horizontal (Fig. 2.5, direction 10) intra directions are applied to generate Modes 0 and 1. The Modes 2 and 3 fill the predicted block with a single depth value derived from the upper and left neighbor blocks, respectively. Additionally, it is not needed to transmit the encoding block nor extra information when applying the Intra-picture skip.

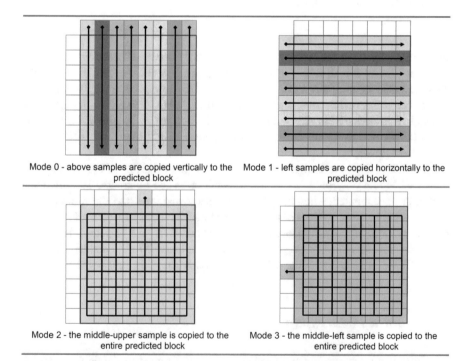

Mode 0 - above samples are copied vertically to the predicted block

Mode 1 - left samples are copied horizontally to the predicted block

Mode 2 - the middle-upper sample is copied to the entire predicted block

Mode 3 - the middle-left sample is copied to the entire predicted block

**Fig. 2.14** Example of an 8 × 8 block encoded with Modes 0, 1, 2 and 3 of the Intra-picture skip

## 2.3   Inter-Frame and Inter-View Predictions

Figure 2.15 shows the search process composed by the Motion Estimation (ME) and Disparity Estimation (DE). ME explores the temporal redundancies among frames at different times, while DE searches for the redundancies among views, i.e., inside of the access units.

Figure 2.16 displays that ME and DE are applied in the prediction of depth maps using a search algorithm to detect the block that contains the highest similarity to the

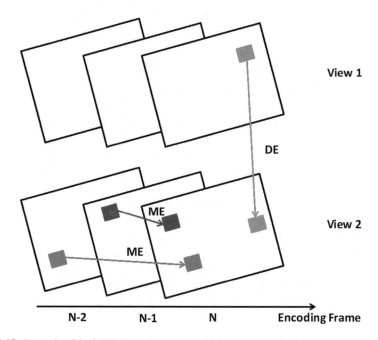

**Fig. 2.15**  Example of the ME/DE search process, which are referencing blocks from other frames

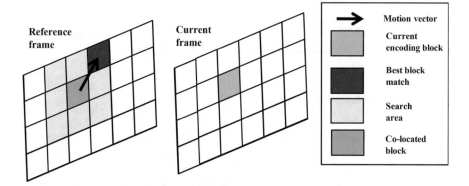

**Fig. 2.16**  Example of the ME/DE search process

encoding block inside a search area in the reference frame. It is performed, mapping the movement/disparity from the current encoding block in a reference frame employing a 2D vector. This vector points to the most similar block in the reference frame by applying a search algorithm, which has a considerable encoding effort. Therefore, the search process is a critical task inside the encoders of the block-based videos [12].

3D-HTM uses Test Zone Search (TZS) as the ME/DE search algorithm and SAD as the similarity criterion. The TZS algorithm can obtain a near optimal performance (i.e., TZS can achieve results quite similar to an exhaustive approach) concerning the quality of the search process. TZS applies an Initial Search Point (ISP) decision and then conducts an iterative search around the best ISP.

TZS selects the best ISP comparing the Sum of Absolute Differences (SAD) results of the: (i) the co-located block (i.e., the block in the same place in the reference frame) vector; (ii) median vector of the encoding block neighborhood (previously encoded blocks around of the current block); and (iii) vector of the largest block for the current CU [13]. Furthermore, TZS can use different patterns in the search process, such as expansive diamond, raster search, and refinement step [13]. The ISP technique combined with these search engines allows reducing around of 23 times the Full Search (FS) (i.e., the exhaustive approach) encoding effort without meaningful impact on the encoded video quality [13]. Even so, TZS still needs a high number of SAD evaluations when compared to other logically equivalent but faster algorithms.

The inter-frame/view prediction also supports the usage of Merge/Skip Mode (MSM), where the encoder derives information of spatial or temporal neighbor blocks. The Merge mode uses a merge list containing the neighbor blocks, whose index is used during the encoding process. As a particular case of MSM, the Skip mode is applied without needing residual information, decreasing the final stream size significantly. Therefore, the encoder utilizes one of the following information for every inter-frame PU: (i) explicit encoding of the motion parameters, (ii) motion Merge mode, or (iii) Skip mode.

## 2.4   Common Encoding Steps

TQ, DC-only, and EC are extra tools used to finish depth map encoding after the execution of the prediction steps described in Sects. 2.2 and 2.3.

### 2.4.1   Transform-Quantization (TQ)

The 3D-HEVC TQ flow for depth maps was inherited from the HEVC texture coding; this flow divides the CU in a quadtree of Transform Unities (TUs) [14], which allows to TQ to range from $4 \times 4$ to $32 \times 32$ samples. Both, a Discrete Cosine Transform (DCT) and a quantization module process the residual information of the predicted block under evaluation. Therefore, when encoding texture, less relevant frequencies to the human vision are attenuated [15], reducing quality losses in the coding process. A Quantization Parameter (QP) defines the quantization strength, and as bigger is the QP as higher is the compression rate, but also as higher is the image quality degradation [15].

## 2.4.2   DC-Only Coding

As an alternative to the TQ flow, 3D-HEVC uses DC-only to obtain higher bitrate gains in smooth regions [7] for block sizes ranging from $8 \times 8$ to $64 \times 64$, where only square blocks are allowed. Instead of using the residual information of depth samples, DC-only considers that a single DC residue encodes the HEVC intra-frame, inter-frame, and inter-view predictions, and considers that two DC residues encode the bipartition modes. The DC residue generation requires to compute a predicted DC (dcp), the original DC (dco), and then to perform a subtraction between them.

It uses the average value of the original block for HEVC intra- and inter-frame predictions, and the average value of each region for the bipartition modes to calculate dco. The use of the average value of the four corners samples of the predicted block for intra-frame prediction generates dcp. Additionally, dcp uses the predicted CPV, which was obtained by applying the algorithm described in Subsection 2.2.2 for the bipartition modes.

## 2.4.3   Entropy Coding

The Entropy Coding (EC) of the depth maps was also inherited from the conventional HEVC texture encoding. Context Adaptive Binary Arithmetic Coding (CABAC) [16] is the primary tool used, which provides a meaningful bitstream reduction by performing lossless entropy compression in the series of syntax elements delivered by the previous encoding tools. Thus, all information produced by the previous tools is processed in EC. The EC results allow determining the RD-cost of all possible modes, enabling to compare those results and to select the one employed in the current block encoding.

## 2.5   Evaluation Methodology – 3D-HEVC Common Test Conditions (CTC)

JCT-3V experts developed the CTC [17] to enable a fair comparison among algorithms of 3D video coding. CTC incorporates eight video sequences that should be encoded in four quantization scenarios. Table 2.2 presents a summary of the characteristics of these video sequences.

**Table 2.2** Details of the CTC video sequences

| Video sequence | Resolution | Encoded frames | Environment | Level of details | Depth structure |
|---|---|---|---|---|---|
| Balloons | 1024 × 768 | 300 | Inside a room | Medium | Medium |
| Kendo | 1024 × 768 | 300 | Inside a room | High | Medium |
| Newspaper_CC | 1024 × 768 | 300 | Inside a room | High | Medium |
| GT_Fly | 1920 × 1088 | 250 | Computer Graphics | Medium | Complex |
| Poznan_Hall2 | 1920 × 1088 | 200 | Inside a room | Medium | Complex |
| Poznan_Street | 1920 × 1088 | 250 | Outside | High | Complex |
| Undo_Dancer | 1920 × 1088 | 250 | Computer Graphics | High | Simple |
| Shark | 1920 × 1088 | 300 | Computer Graphics | Medium | Complex |

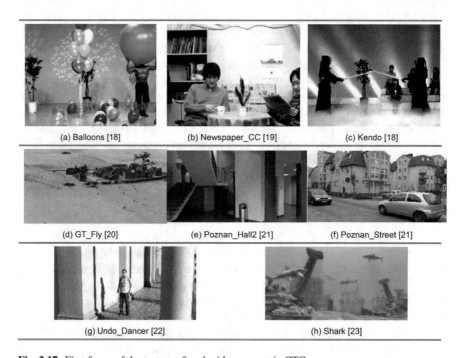

(a) Balloons [18]      (b) Newspaper_CC [19]      (c) Kendo [18]

(d) GT_Fly [20]      (e) Poznan_Hall2 [21]      (f) Poznan_Street [21]

(g) Undo_Dancer [22]      (h) Shark [23]

**Fig. 2.17** First frame of the texture of each video present in CTC

Each video sequence covers a vast quantity of specific characteristics, such as high movement, transparency, details and camera noise. Figures 2.17 and 2.18 present the first texture frame and the depth map of the central view, respectively. The following set of videos should be evaluated in CTCs: (a) Balloons [18], (b) Newspaper_CC [19], (c) Kendo [18], (d) GT_Fly [20], (e) Poznan_Hall2 [21], (f) Poznan_Street [21], (g) Undo_Dancer [22], and (h) Shark [23].

The experts designed the 3D-HTM [4], which is a 3D-HEVC reference software that includes an implementation of the 3D-HEVC encoder used to evaluate the CTC video sequences. 3D-HTM produces relevant information, such as texture encoding

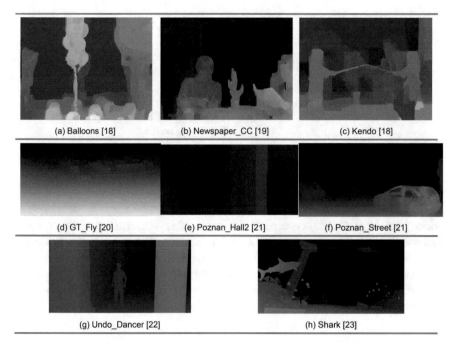

(a) Balloons [18]    (b) Newspaper_CC [19]    (c) Kendo [18]

(d) GT_Fly [20]    (e) Poznan_Hall2 [21]    (f) Poznan_Street [21]

(g) Undo_Dancer [22]    (h) Shark [23]

**Fig. 2.18** First frame of the depth maps of each video available in CTC

time, encoding time of depth maps, bitrate for each encoded view, and the resulting Peak Signal-to-Noise Ratio (PSNR); i.e., the subjective video quality of each view. After encoding a video sequence with 3D-HTM, it is possible to use another module of this reference software containing a 3D-HEVC decoder. This module allows decoding the views and, posteriorly, generating an intermediary view by employing a view synthesis interpolator. Consequently, one can assess the objective quality of the synthesized view and compare it with a related work fairly.

3D-HTM uses four QP pairs (QP-texture, QP-depth) for each video sequence: (25, 34), (30, 39), (35, 42), and (40, 45) to follow the CTC definition, where the QP represents the quantization impact in the encoding process. The higher the QP value, the higher are the reductions in the video quality, also resulting in a bitrate reduction.

The evaluation of these four QP pairs allows drawing Rate-Distortion (RD) curves to compare and analyze encoding algorithms. Moreover, G. Bjontegaard [24] proposed the Bjontegaard Delta-rate (BD-rate) technique, which is a metric that provides the bitrate reduction for similar video quality. The BD-rate is computed by interpolating the resulting Peak Signal-to-Noise Ratio (PSNR) and the bitrate obtained in one experiment and comparing this result with another implementation. However, in the depth maps coding, the PSNR of the encoded view of the depth map does not provide relevant information for analysis; consequently, the BD-rate of the depth maps are not relevant. Therefore, for evaluating the BD-rate in Multiview

**Table 2.3** Configurations used in the experiments of this Book

| Profile | All intra | Random access |
|---|---|---|
| GOP size | 1 | 8 |
| Intra-Period | 1 | 24 |
| Bits per pixel | 8 | |
| Texture/Depth views | 3/3 | |
| QP-pairs | (25, 34), (30, 39), (35, 42), (40, 45) | |
| DMM Evaluation | Enabled | |
| Depth Intra Skip | Enabled | |
| VSO | Enabled | |
| Rate Control | Disabled | |
| 3D-HTM version | 16.0 | |

Video plus Depth (MVD) systems, it is necessary to synthesize intermediary views and use their PSNR as the reference for this evaluation. Therefore, in this Book, the evaluated BD-rate is obtained using the average PSNR of the synthesized views and the total bitrate; i.e., the sum of the bitrate of all texture views and depth maps.

Table 2.3 displays the principal configurations applied to all the experiments of this Book. These configurations are following the CTC definitions.

# References

1. Saldanha, M., G. Sanchez, C. Marcon, and L. Agostini. 2019. Fast 3D-HEVC depth maps encoding using machine learning. *IEEE Transactions on Circuits and Systems for Video Technology* 1–1: 12.
2. Marpe, D., H. Schwarz, S. Bosse, B. Bross, P. Helle, T. Hinz, H. Kirchhoffer, H. Lakshman, T. Nguyen, S. Oudin, et al. 2010. Video compression using nested quadtree structures, leaf merging, and improved techniques for motion representation and entropy coding. *IEEE Transactions on Circuits and Systems for Video Technology* 20–12: 1676–1687.
3. Sullivan, G.J., J.-R. Ohm, W.-J. Han, T. Wiegand, et al. 2012. Overview of the high efficiency video coding (HEVC) standard. *IEEE Transactions on circuits and systems for video technology* 22 (12): 1649–1668.
4. ITU-T VCEG and ISO/IEC MPEG. 2017. 3D-HEVC Test Model. Source: https://hevc.hhi. fraunhofer.de/svn/svn_3DVCSoftware/tags/HTM-16.0/, Sep 2017.
5. Merkle, P., K. Müller, D. Marpe, and T. Wiegand. 2016. Depth intra coding for 3D video based on geometric primitives. *IEEE Transactions on Circuits and Systems for Video Technology* 26–3: 570–582.
6. Lee, J., M. Park, and C. Kim. 2015. 3d-ce1: depth intra skip (dis) mode, Technical Report, ISO/IEC JTC1/SC29/WG11, 5.
7. Liu, H., and Y. Chen. 2014. Generic segment-wise DC for 3D-HEVC depth intra coding. *IEEE International Conference on Image Processing*: 3219–3222.
8. Lainema, J., F. Bossen, W.-J. Han, J. Min, and K. Ugur. 2012. Intra coding of the HEVC standard. *IEEE Transactions on Circuits and Systems for Video Technology* 22 (12): 1792–1801.
9. Zhao, L., L. Zhang, S. Ma, and D. Zhao. 2011. Fast mode decision algorithm for intra prediction in HEVC. *IEEE Visual Communications and Image Processing*: 1–4.
10. Tech, G., H. Schwarz, K. Müller, and T. Wiegand. 2012. 3D video coding using the synthesized view distortion change. *Picture Coding Symposium*: 25–28.

11. Müller, K., H. Schwarz, D. Marpe, C. Bartnik, S. Bosse, H. Brust, T. Hinz, H. Lakshman, P. Merkle, F.H. Rhee, et al. 2013. 3D high-efficiency video coding for multi-view vídeo and depth data. *IEEE Transactions on Image Processing* 22 (9): 3366–3378.
12. Cheng, Y.-S., Z.-Y. Chen, and P.-C. Chang. 2009. An H.264 spatio-temporal hierarchical fast motion estimation algorithm for high-definition video. *IEEE International Symposium on Circuits and Systems*: 880–883.
13. Tang, X.-l., S.-K. Dai, and C.-H. Cai. 2010. An analysis of TZSearch algorithm in JMVC. *International Conference on Green Circuits and Systems*: 516–520.
14. Winken, M., P. Helle, D. Marpe, H. Schwarz, and T. Wiegand. 2011. Transform coding in the HEVC test model. *IEEE International Conference on Image Processing*: 3693–3696.
15. Budagavi, M., A. Fuldseth, and G. Bjontegaard. 2014. *High efficiency video coding (HEVC): Algorithms and architectures*. Vol. 6, 141–169. Cambridge: Springer.
16. Marpe, D., H. Schwarz, and T. Wiegand. 2003. Context-based adaptive binary arithmetic coding in the H. 264/AVC video compression standard. *IEEE Transactions on circuits and systems for video technology* 13–7: 620–636.
17. Müller. K, and A. Vetro. 2014. Common test conditions of 3DV Core experiments, Technical Report, ISO/IEC JTC1/SC29/WG11, 7p.
18. Tanimoto Lab NICT. 2017. National Institute of Information and Communication Technology. Source: http://www.tanimoto.nuee.nagoya-u.ac.jp/, Sep 2017.
19. Ho, Y.-S., E.-K. Lee, C. Lee. 2008. M15419, multiview video test sequence and camera parameters, Technical Report, ISO/IEC JTC1/SC29/WG11, 6p.
20. Zhang, J.; Li, R.; Li, H.; Rusanovskyy, D.; Hannuksela, M. M.. 2011. Ghost Town Fly 3DV sequence for purposes of 3DV standardization, Technical Report, ISO/IECJTC1/SC29/WG11, 5p.
21. Domañski, M., T. Grajek, K. Klimaszewski, M. Kurc, O. Stankiewicz, J. Stankowski, and K. Wegner. 2009. Poznan multiview video test sequences and camera parameters, Technical Report, ISO/IEC JTC1/SC29/WG11, 6.
22. Rusanovskyy D., P. Aflaki, and M. Hannuksela. 2011. Undo Dancer 3DV sequence for purposes of 3DV standardization, Technical Report, ISO/IEC JTC1/SC29/WG11, 6.
23. NICT. 2017. National Institute of Information and Communication Technology. Source: ftp://ftp.merl.com/pub/tian/NICT-3D/Shark/.
24. Bjontegaard, G. 2001. Calculation of average PSNR differences between RD-curves, Technical Report, ITU-T SC16/SG16 VCEG-M33, 4.
25. Chen, Y, G. Tech, K. Wegner, and S. Yea. 2015. Test model 11 of 3D-HEVC and MV-HEVC, Technical Report, ISO/IEC JTC1/SC29/WG11, 58.

# Chapter 3
# State-of-the-Art Overview

This chapter describes works targeting four topics of depth map coding. Sections 3.1 and 3.2 present and discuss the works targeting to reduce the computational effort of the intra-frame and inter-frame/view predictions, respectively. The time-saving algorithms were classified in (i) mode-level, (ii) block-level, and (iii) quadtree-level. The timesaving techniques of the mode-level focus on simplifying a single encoding mode. The block-level algorithms skip the complete evaluation of some encoding steps if a given criterion is met. For instance, if a block and its neighbor block are highly correlated with the current encoding block (i.e., highly similar characteristics), then the current block can be encoded using the same mode of the neighbor block, skipping the remaining evaluations. The quadtree-level algorithms limit the minimum and maximum quadtree-level adaptively according to the encoding block characteristics. For example, there is no reason to evaluate small block sizes if the encoding block is nearly homogeneous since a higher block size can represent almost the same information without decreasing the encoding quality. Besides, coding an area with many details is more difficult when using large block sizes.

## 3.1 State-of-the-Art Concerning the Reduction of the Computational Effort Applied to the Intra-Frame Prediction

There are several works presenting solutions for reducing the computational effort on the depth map intra-frame prediction of the 3D-HEVC. Sections 3.1.1, 3.1.2, and 3.1.3 discuss the works employing mode-, block-, and quadtree-level decision algorithms, respectively.

© Springer Nature Switzerland AG 2020
G. Sanchez et al., *Algorithms for Efficient and Fast 3D-HEVC Depth Map Encoding*, https://doi.org/10.1007/978-3-030-25927-3_3

### 3.1.1   State-of-the-Art of the Mode-Level Decision Algorithms

Regarding mode-level decision algorithms, some works propose to diminish the encoding effort of Rough Mode Decision (RMD) and Most Probable Modes (MPM) selection (e.g., [1, 2]), some other works suggest to decrease the encoding effort of the Transform-Quantization (TQ) and/or Direct Component-only (DC-only) flows (e.g., [2–4]), and others recommend to reduce the wedgelet list evaluations (e.g., [1, 5]). The authors of this Book have also proposed two works [6, 7] focusing on reducing the wedgelet list evaluation.

Zhang et al. [1] proposed to accelerate the RMD and MPM selection, and the Intra_Wedge wedgelet list evaluation. They classify the encoding block in one of three types according to the variance value of neighbor reference samples: (i) smooth, (ii) edge, or (iii) normal. A smooth block requires inserting only the planar mode into the RD-list; if the block is classified as an edge, the algorithm uses the same orientation of the intra-frame direction to decide the modes inserted into the RD-list. Besides, it uses the Intra_Wedge pattern orientation to avoid evaluating the entire Intra_Wedge set. If the block is classified as normal, all modes are evaluated, and the original 3D-HEVC Test Model (3D-HTM) encoding flow is performed. Their work saves 27.9% of the depth map coding time while increases 1.03% the BD-rate.

The work of Zhang et al. [2] performs mode- and block-level (described in Sect. 3.1.2) decisions. The modes 10 (horizontal), 26 (vertical), planar, DC, and bipartition are always inserted into the Rate Distortion-list (RD-list). The mode-level decision also analyses the parental block (a larger block containing the current block) encoding mode. If the parental block is not encoded by the planar, DC or bipartition modes, then the mode-level decision also inserts two extra intra-frame prediction modes into the RD-list. Otherwise, the RD-list are evaluated without inserting more modes on it. The proposed mode-level decision reduces 24.4% of the depth map coding time with an increase of 0.51% in the BD-rate.

Peng, Chiang, and Lie [4] proposed an algorithm for saving time in the TQ and DC-only flows using mode-, block- and quadtree-level decisions. Since its mode- and block-level decisions do not work alone, they are explained together here, while its quadtree-level decision is described in Sect. 3.1.3. Step 1 of the algorithm evaluates only planar, DC, horizontal, vertical, and MPM modes and compares the obtained results with a threshold. The process stops if the threshold criterion is met, without evaluating the remaining HEVC intra-prediction directions and the bipartition modes; else, the algorithm executes the step 2, where bipartition modes are evaluated generating a new Rate Distortion-cost (RD-cost). If the new RD-cost of the bipartition modes is smaller than the previous RD-cost obtained in step 1, then the encoding process ends. On the contrary, the algorithm performs step 3, computing the RD-cost for the remaining intra-frame prediction modes that share a similar direction to the best wedgelet obtained in Intra_Wedge. Again, the obtained RD-cost is compared with the RD-cost obtained in step 1, finishing the process when the new RD-cost is smaller. Otherwise, the original encoding is done evaluating all the remaining modes. This scheme obtained 21.6% of timesaving with a BD-rate increase of 0.5%.

Gu, Zheng, Ling, and Zhang [3] implemented a standard 3D-HTM algorithm for simplifying the DC-only evaluation without changing the TQ encoding flow. The three best-encoded modes in TQ flow are stored during their evaluation. If the best-encoded mode is planar or DC, and the variance of the encoding block is lower than a threshold, then only planar and DC modes are assessed in the DC-only flow. If the best-encoded mode by TQ flow is not planar or DC, then only the three best-encoded modes are evaluated by their RD-cost in the DC-only flow. Since this algorithm is executed by default in 3D-HTM, then its encoding time is considered as the baseline for the remaining evaluations. Nonetheless, before being integrated with the reference software, it was capable of reducing the encoding time in 17.0%, while increasing the BD-rate in 0.46%.

The remaining mode-level decision algorithms focus on reducing the number of wedgelets evaluated by the Intra_Wedge. Our previous work [7] designed a gradient filter in the borders of the block to detect the most promising positions and to evaluate Intra_Wedge patterns. It evaluates only the wedgelets that touch the positions with high gradients values in the borders. Significant advances providing higher wedgelets evaluation skips were designed in our previous work [6] when analyzing the two highest gradients of different borders and evaluating wedgelets that touch near positions. When applying Random Access configuration, the results of encoding timesaving in these evaluations vary between 4.9% and 8.2% with a BD-rate increase between 0.33% and 1.47%.

Fu et al. [5] propose to divide the encoding block into two regions using a wedgelet with one of four slopes passing in the middle of the block: (i) vertical, (ii) horizontal, (iii) 45°, and (iv) 135°. In each evaluation, the variance of each region is computed to verify if each region has a strong potential of containing an edge region. With that analysis, the wedgelet search process is reduced by only evaluating wedgelets with a high potential of being selected. This technique reduced 52.9% of the evaluated wedgelets, with an increase of 0.49% in the BD-rate. Fu et al. [5] do not present any evaluation regarding the timesaving provided by their solution.

### 3.1.2 State-of-the-Art in Block-Level Decision Algorithms

Regarding block-level decision algorithms, the work [3] and our previous work [8] proposed skipping the entire bipartition modes evaluation when the traditional HEVC intra-frame prediction encodes the block effectively. Thus, these works do not make any evaluation in Intra_Wedge and Intra_Contour when the skip is performed. Moreover, some works focus on removing the TQ flow evaluation [2], and on encoding only one flow [9] if a given criterion is met.

The work of Gu et al. [10] calculate the variance of the encoding block and employ a threshold based on QP for performing a fast block-level decision (the algorithm used in the 3D-HTM standard). If the encoding block has a low variance or if planar is the best mode in RD-list inserted by HEVC intra-frame prediction, then the bipartition modes are rarely selected. In this case, Gu et al. [10] do not perform the bipartition modes evaluation. Since it is already executed by default in 3D-HTM, then its encoding time is considered as the baseline for the remaining evaluations. Before its integration with the reference software, it was capable of reducing the encoding time in 26.5% with an increase of 0.02% in BD-rate.

In [8], we designed an approach similar to that described in [10]; however, instead of using the variance of the encoding block, it uses a Simplified Edge Detector (SED), which compares the highest difference among the four corners of the block with a threshold decided according to the block size and video resolution. If the value is below the threshold, then the encoding block represents a nearly constant region, and the bipartition modes evaluation is skipped. When assessed in random access configuration, it reduced 0.064% the BD-rate and 5.9% the encoding time considering the entire encoder and 23.8% when considering only depth map encoding time.

Zhang et al. [2] analyze the correlation between the parental Prediction Unit (PU) and the children PUs (i.e., smaller blocks inside the given PU). They concluded that when the best result of the parental PU was obtained using DC-only, more than 94% of its child PUs are also encoded with DC-only. Therefore, Zhang et al. [2] designed a block-level algorithm to skip only the TQ flow in the smaller blocks evaluation if the current block parent is DC-only coded. With this decision integrated with their mode-level described in Sect. 3.1.1, they reached 32.87% of time reduction in depth map encoding with a BD-rate increase of 0.54%.

Conceição et al. [9] propose to skip the evaluation of the remaining encoding modes according to the RD-cost obtained by Skip and Intra-picture skip modes. The main idea is analyzing the RD-cost obtained in these modes as a decision to avoid the computation of the remaining encoding modes. Since the proposed technique requires working with Skip and Intra-picture skip modes together, then it is better described in Sect. 3.2 because Skip mode is not applied in All Intra configuration. Besides, the work does not mention how intra encoding frames are evaluated.

### 3.1.3   State-of-the-Art in Quadtree-Level Decision Algorithms

Peng et al. [4] introduced a quadtree-level decision algorithm that uses the variance of the current block. If the variance is higher than a threshold that is defined according to the encoding QP, then the algorithm divides the quadtree into smaller blocks; otherwise, it computes the variances of the four sub-blocks inside the encoding block. The algorithm finishes the quadtree expansion only when the highest variance in these sub-blocks are lower than the current block variance. Working along with the mode decision explained in Sect. 3.1.1, the solution designed by Peng et al. [4] reduced 37.6% the encoding time with 0.8% of BD-rate increase.

H. Zhang et al. [11] proposed a QP-based depth quadtree limit for splitting the quadtree only when the encoding block carries details that are not well encoded by large blocks. Additionally, the authors use a classical corner detection algorithm [12] to identify if the information is relevant for smaller block sizes. Their work reduces 41% the encoding time with an increase of 0.44% in the BD-rate.

H. Chen et al. [13] designed an algorithm for early termination of the intra CU splitting. When the algorithm divides a CU into four new sub-CUs, and the first sub-CU selects Intra-picture skip as the best encoding mode, the RD-cost obtained in the parental CU is compared to the RD-cost obtained by Intra-picture skip mode. If the RD-cost of the parental CU is smaller, then the parental CU is selected in the encoding process, and further evaluations are not required. Otherwise, the remaining sub-CUs are evaluated, and the sum of all sub-CU costs is compared with the parental CU cost. Again, in case the parental CU cost is smaller, then it is selected as the best partition. This approach allows reducing 45.1% of the depth map encoding effort with a BD-rate increase of 0.09%.

## 3.1.4 Summary of the State-of-the-Art in Intra-Frame Prediction

Table 3.1 presents a summary of the related works; the focused tool is highlighted for mode-level, the skipped tools are highlighted for block-level, and it is detached the quadtree-level decision. Among the related works, one can see different characteristics that were explored in depth map coding. In mode-level decisions, Intra_Wedge is the most focused tool due to its high encoding effort associated with it. In block-level decisions, skipping bipartition modes was the most focused encoding tool. Besides, only a few works focused on exploring quadtree-level decisions for depth map coding; however, the highest time-savings were achieved in these works.

**Table 3.1** Summary of state-of-the-art in intra-frame prediction

| Work | Tool focused on mode-level | | | | Tool skipped on block-level | | | | Quadtree decision |
| | Intra_Wedge | RMD/MPM | TQ | DC-only | Bipartition modes | RMD/MPM | TQ | DC-only | |
|---|---|---|---|---|---|---|---|---|---|
| H.-B. Zhang et al. [1] | X | X | | | | | | | |
| H.-B. Zhang et al. [2] | | X | X | X | | X | | | |
| Z. Gu et al. [3] | | | | X | | | | | |
| K.-K Peng et al. [4] | | X | X | X | X | | | | X |
| C.-H. Fu et al. [5] | X | | | | | | | | |
| M. Saldanha et al. [6][a] | X | | | | | | | | |
| G. Sanchez et al. [7][a] | X | | | | | | | | |
| G. Sanchez et al. [8][a] | | | | | X | | | | |
| R. Conceição et al. [9] | | | | | X | X | X | X | |
| Z. Gu et al. [10] | | | | | X | | | | |
| H. Chen et al. [13] | | | | | | | | | X |
| H.-B. Zhang et al. [11] | | | | | | | | | X |

[a]our previous works

## 3.2   State-of-the-Art in Reducing the Computational Effort on the Inter-Frame/View Prediction

A few works are found in the literature reducing the encoding effort on inter-frame/ view prediction of depth maps because most of the encoding tools were inherited from the texture. These works are analyzed in three classes: (i) mode-level, which focuses on fast Motion Estimation (ME) / Disparity Estimation (DE) algorithms; (ii) block-level, which avoids computing all available modes for a given block; and (iii) quadtree-level, which focuses on pruning the quadtree expansion.

Afonso et al. [14] designed an algorithm for reducing the encoding effort of the DE. Their algorithm explores the 3D-HEVC horizontal camera arrangement by performing a fast horizontal-only DE search. This search is performed in three steps, where the first step uses a high subsampling for the best block match horizontally. Then, the next steps refine the previous step result by performing smaller subsampling evaluations around the best block match obtained in the previous step. It achieved a reduction between 32.7% and 61.8% in the Sum of Absolute Differences (SAD) computation with an increase in the BD-rate between 0.31% and 0.48%. The authors do not present specific timesaving results.

As previously mentioned in Sect. 3.1.2, Conceição et al. [9] proposed a block-level decision scheme that exploits the usage of Skip and Intra-picture skip for the 3D-HEVC depth map encoding. Since these are the most used modes, when the RD-costs obtained encoding them are low, it is high the probability of select them as the final decision. Therefore, the RD-cost was used as a decision for removing the remaining modes evaluation. They provided a time reduction between 24.4% and 33.7% in the depth map encoding with an increase of BD-rate between 0.067% and 0.409%. This was the only work found in the literature proposing block-level decisions for random access modes.

Two works focus on inter-view/frame encoding at the quadtree-level [15, 16]. Mora, Jung, Cagnazzo and Pesquet-Popescu [16] propose to use inter-component for reducing the encoding effort spent in the depth map coding by limiting the quadtree expansion based on the Coding Unit (CU) partition information of the col-located texture CU. Considering the texture and depth maps are captured from the same viewpoint at the same time, then the quadtrees of both are highly correlated. Thus, when encoding P- and B-frames, the quadtree expansion of the depth maps is limited to the correlated texture quadtree. The 3D-HTM reference software includes this solution, which allows reducing 52% the encoding time with an increase of 1.024% in the BD-rate.

Lei et al. [15] designed a fast mode decision algorithm for depth map coding based on the grayscale similarities and the inter-view correlation. The mode decision adopts an early CU termination and an early PU mode decision for encoding dependent views. In the CU termination decision, the encoder employs a threshold for classifying the grayscale similarities between the current encoding CU and the co-located block in the reference frame. When the similarity is high, the quadtree depth level of the current CU is constrained by the level of the co-located quadtree in the reference frame. Additionally, a grayscale similarity and an inter-view correlation are applied in the PU mode decision. The PU mode decision concludes when the co-located CU in the reference view selects merge or inter 2 N × 2 N as the best mode. It reached an encoding time reduction of 41.5% with 2.06% increase in the BD-rate.

Only a few works are focusing on inter-frame/view predictions in the literature, where only one was focused on mode-level decisions, one focused on block-level decisions, and two focused on quadtree-level decisions.

# References

1. Zhang, H.-B., C.-H. Fu, Y.-L. Chan, S.-H. Tsang, and W.-C. Siu. 2015. Efficient depth intra mode decision by reference pixels classification in 3D-HEVC. *IEEE International Conference on Image Processing* 2015: 961–965.
2. Zhang, H.-B.; S.-H. Tsang, Y.-L. Chan, C.-H. Fu, and W.-M. Su. 2015. Early determination of intra mode and segment-wise DC coding for depth map based on hierarchical coding structure in 3D-HEVC. In: Asia-Pacific Signal and Information Processing Association Annual Summit and Conference, 374–378.
3. Gu, Z., J. Zheng, N. Ling, and P. Zhang. 2015. Fast segment-wise DC coding for 3D video compression. *IEEE International Symposium on Circuits and Systems* 2015: 2780–2783.
4. Peng, K.-K., J.-C. Chiang, and W.-N. Lie. 2016. Low complexity depth intra coding combining fast intra mode and fast CU size decision in 3D-HEVC. *IEEE International Conference on Image Processing*: 1126–1130.
5. Fu, C.-H., H.-B. Zhang, W.-M. Su, S.-H. Tsang, and Y.-L. Chan. 2015. Fast wedgelet pattern decision for DMM in 3D-HEVC. *IEEE International Conference on Digital Signal Processing*: 477–481.
6. Saldanha, M., B. Zatt, M. Porto, L. Agostini, and G. Sanchez. 2016. Solutions for DMM-1 complexity reduction in 3D-HEVC based on gradient calculation. In: IEEE 7th Latin American Symposium on Circuits & Systems, 211–214.
7. Sanchez, G., M. Saldanha, G. Balota, B. Zatt, M. Porto, and L. Agostini. 2014. A complexity reduction algorithm for depth maps intra prediction on the 3D-HEVC. In: IEEE Visual Communications and Image Processing Conference, 137–140.
8. ———. 2014. Complexity reduction for 3D-HEVC depth maps intra-frame prediction using simplified edge detector algorithm. In: IEEE International Conference on Image Processing, 3209–3213.
9. Conceição, R., G. Avila, G. Corrêa, M. Porto, B. Zatt, and L. Agostini. 2016. Complexity reduction for 3D-HEVC depth map coding based on early skip and early DIS scheme. *IEEE International Conference on Image Processing* 2016: 1116–1120.
10. Gu, Z., J. Zheng, N. Ling, and P. Zhang. 2013. Fast intra prediction mode selection for intra depth map coding, Technical Report, ISO/IEC JTC1/SC29/WG11, 4.
11. Zhang, H.-B., Y.-L. Chan, C.-H. Fu, S.-H. Tsang, and W.-C. Siu. 2016. Quadtree decision for depth intra coding in 3D-HEVC by good feature. *IEEE International Conference on Acoustics, Speech and Signal Processing*: 1481–1485.
12. Harris, C., and M. Stephens. 1988. A combined corner and edge detector. In: Alvey vision conference, 147–151.
13. Chen, H., C.-H. Fu, Y.-L. Chan, and X. Zhu. 2018. Early intra block partition decision for depth maps in 3D-HEVC. *IEEE International Conference on Image Processing*: 1777–1781.
14. Afonso, V., A. Susin, M. Perleberg, R. Conceição, G. Corrêa, L. Agostini, B. Zatt, and M. Porto. 2018. Hardware-friendly unidirectional disparity-search algorithm for 3D-HEVC. *IEEE International Symposium on Circuits and Systems*: 1–5.

15. Lei, J., J. Duan, F. Wu, N. Ling, and C. Hou. 2018. Fast mode decision based on grayscale similarity and inter-view correlation for depth map coding in 3D-HEVC. *IEEE Transactions on Circuits and Systems for Video Technology* 28 (3): 706–718.
16. Mora, E.G., J. Jung, M. Cagnazzo, and B. Pesquet-Popescu. 2014. Initialization, limitation, and predictive coding of the depth and texture quadtree in 3D-HEVC. *IEEE Transactions on Circuits and Systems for Video Technology* 24 (9): 1554–1565.

# Chapter 4
# Encoding Time Reduction for 3D-HEVC Intra-Frame Prediction of Depth Maps

This chapter presents researches related to encoding time reduction for intra-frame prediction. Section 4.1 investigates the encoding time and mode distribution for intra-frame prediction of depth maps. These investigations are contributions published in previous works of the Book's authors [1, 2]. Section 4.2 describes algorithms designed for reducing the encoding time of depth map intra-frame prediction, and Sect. 4.3 shows the experimental results and discussion about the proposals presented in this Section.

## 4.1 Performance Analysis

Two subsections with complementary analysis encompass Sect. 4.1. The first one (4.1.1) discusses the time allocation of each encoding mode in four quantization scenarios. The second one (4.1.2) presents the encoding mode distribution of all block sizes and four Quantization Parameters (QPs). The videos were evaluated under all-intra-frame scenario using 3D-HEVC Test Model (3D-HTM) 16.0 and the configurations described in Sect. 2.5, which already include by default the techniques proposed by the works [3], for avoiding some evaluations of the bipartition modes, and [4], for reducing the encoding time of the DC-only evaluation. All experiments described in this section were conducted in a server containing two Xeon E5–2660 processors with 96 GB of main memory.

### 4.1.1 Encoding Time Distribution

The experimental results presented in Table 4.1 shows the percentage of the encoding time distribution for each component per video and per QP-pair. Additionally, Table 4.1 displays the average time for encoding one texture frame and its

© Springer Nature Switzerland AG 2020

G. Sanchez et al., *Algorithms for Efficient and Fast 3D-HEVC Depth Map Encoding*, https://doi.org/10.1007/978-3-030-25927-3_4

**Table 4.1** Encoding time evaluation for texture and depth map coding using the 3D-HEVC all intra configuration and four QP-pairs (25/34, 30/39, 35/42, 40/45)

| Video | Texture (%) | | | | Depth (%) | | | | Total (s) | | | |
|---|---|---|---|---|---|---|---|---|---|---|---|---|
| | 25/34 | 30/39 | 35/42 | 40/45 | 25/34 | 30/39 | 35/42 | 40/45 | 25/34 | 30/39 | 35/42 | 40/45 |
| Balloons | 15.7 | 14.4 | 14.9 | 14.5 | 84.3 | 85.6 | 85.1 | 85.5 | 33.72 | 34.78 | 33.64 | 34.48 |
| Kendo | 15.9 | 15.2 | 15.4 | 13.4 | 84.1 | 84.8 | 84.6 | 86.6 | 31.62 | 32.90 | 32.52 | 32.77 |
| Newspaper | 13.2 | 11.0 | 11.7 | 11.8 | 86.8 | 89.0 | 88.3 | 88.2 | 45.62 | 45.53 | 42.66 | 42.21 |
| GT_Fly | 14.8 | 13.3 | 14.5 | 14.8 | 85.2 | 86.7 | 85.5 | 85.2 | 83.11 | 82.58 | 70.58 | 67.04 |
| Poznan_Hall2 | 18.2 | 16.2 | 16.8 | 15.6 | 81.8 | 83.8 | 83.2 | 84.4 | 47.67 | 49.60 | 47.72 | 49.20 |
| Poznan_Street | 13.6 | 12.2 | 13.7 | 13.5 | 86.4 | 87.8 | 86.3 | 86.5 | 95.89 | 93.01 | 77.32 | 73.96 |
| Undo_Dancer | 18.0 | 16.7 | 16.0 | 14.8 | 82.0 | 83.3 | 84.0 | 85.2 | 78.98 | 73.53 | 68.47 | 67.37 |
| Shark | 16.1 | 14.7 | 15.2 | 14.5 | 83.9 | 85.3 | 84.8 | 85.5 | 87.93 | 88.99 | 80.80 | 80.22 |
| Average | 15.7 | 14.2 | 14.8 | 14.1 | 84.3 | 85.8 | 85.2 | 85.9 | 63.07 | 62.62 | 56.71 | 52.43 |

associated depth map. The encoding time varies according to the video characteristics; however, on average, the encoding time decreases with the increase of QP.

The time spent during the texture coding ranges from 11.0% to 18.2%, which is much lower than the encoding time of the depth maps. On average, depth map coding is 5.8 times more time-consuming than the texture coding in the all-intra-frame scenario. It occurs in all-intra configuration because texture coding only employs traditional HEVC intra-frame prediction following TQ flow, while depth map coding still uses new encoding tools such as the Intra_Wedge, Intra_Contour, Intra-picture skip, and DC-only evaluations.

Two new evaluations of encoding time for depth maps complement this experiment: (i) one describing the percentage of encoding time used per block size regarding the QP-depth, and (ii) another one detailing the encoding time behavior inside a given block size. Both experiments evaluate all video sequences and frames setting the encoder parameters according to the Common Test Conditions (CTC) and the configuration described in Sect. 2.5. Figure 4.1 illustrates the first evaluation. The encoding time distribution represents the percentage of time required for depth map encoding for a given scenario.

The QP-depth variation gives an encoding time distribution with slight differences because of the applied QP-based timesaving algorithms ([3, 4]). However, it is expected that the selection of different block sizes varies according to the QP-depth, as discussed in Sect. 4.1.2, and it can bring several advantages when designing a timesaving solution.

It is vital to distinguish the time spent on each encoding tool. Since some modes share encoding steps (e.g., the Rough Mode Decision (RMD) and Most Probable Modes (MPM) selection in HEVC intra-frame prediction are used for both TQ and

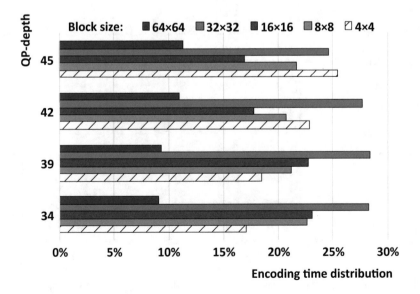

**Fig. 4.1** Encoding time distribution of depth maps according to the block size and QP-depth

DC-only flows), the subsequent analysis focuses on the encoding steps instead of the encoding tools. The following steps were considered in this new evaluation: (i) RMD and MPM selection; (ii) Intra_Wedge wedgelet search; (iii) Intra_Contour pattern generation; (iv) Rate Distortion-list (RD-list) evaluation in TQ flow; (v) RD-list assessment in DC-only flow; and (vi) Intra-picture skip evaluation. Moreover, steps (iv) to (vi) consider the time spent in entropy coding to generate the Rate Distortion-cost (RD-cost). Figure 4.2 displays the encoding time distribution among the intra-frame prediction steps for each block size, regarding the corners of QP-depth values, which complements the previous experiment.

The Intra-picture skip and DC-only evaluations are not applied to 4 × 4 blocks because these modes are used only in block sizes ranging from 8 × 8 to 64 × 64. Additionally, Intra_Wedge and Intra_Contour are not present in 64 × 64 blocks because the bipartition modes are only employed on block sizes ranging from 4 × 4 to 32 × 32.

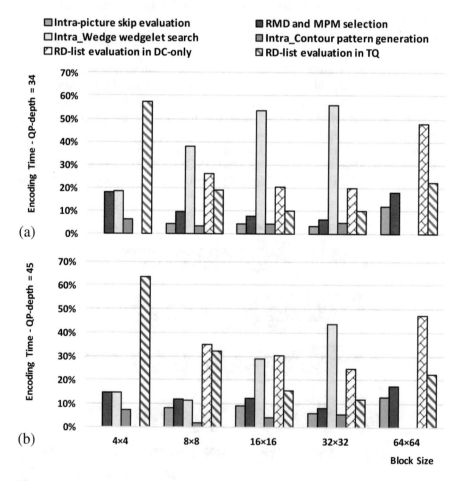

**Fig. 4.2** Distribution of the encoding time considering six encoding steps with (**a**) QP-depth = 34 and (**b**) QP-depth = 45

The RD-list evaluation in TQ flow is the most time-consuming operation for both evaluated QP-depths when considering 4 × 4 blocks, where Intra-picture skip and DC-only evaluations are not available. Intra_Wedge displays the second highest percentage of encoding time. The RMD and MPM selection and Intra_Contour pattern generation also require a significant amount of time to be processed when compared to the other block sizes.

The 8 × 8 block size shows a significant variation on the time distribution of the encoding tools for different values of QP-depth. For lower QP-depths, the Intra_Wedge wedgelet search represents the highest percentage of time among the encoding steps. This behavior changes entirely for high QP-depths (where the RD-list evaluation in DC-only and TQ have the highest percentage of time demanded) because the bipartition modes were proposed to handle with depth map details, intending to preserve the borders, as previously discussed. Applying the heuristic proposed by [3] reduces the effort spent in the bipartition modes since higher QP-depths reduce the image details.

The wedgelet search in Intra_Wedge and the RD-list evaluation in DC-only flows present the first and second positions in the percentage of time spent to encode 16 × 16 and 32 × 32 block sizes for the evaluated QP-depths, respectively. The Intra_Wedge evaluation needs a higher percentage of time for higher block sizes because the solution introduced in [3], which is executed by default in 3D-HTM, skips the bipartition modes evaluation for low variance blocks. Since larger blocks tend to have higher variance, then fewer skips of bipartition modes happen. The RD-list evaluation in DC-only is the most time-consuming operation for processing 64 × 64 blocks; it signifies almost 50% of the total encoding time for both QP-depths. The DC-only execution time is higher for bigger block sizes because DC-only was designed to be used in homogeneous blocks or blocks that can be divided into two homogeneous areas. Thus, the heuristic proposed by [4] tends to skip the DC-only evaluation for heterogeneous regions.

Intra_Contour and Intra-picture skip have low execution times for almost all block sizes. The Intra-picture skip encoding time is only higher than 10% for the 64 × 64 encoding blocks.

Essential conclusions are also obtained considering the time distribution of the encoding steps versus QP-depth values. For instance, the percentage of time spent in the Intra_Wedge calculation is inversely proportional to the QP-depth values used for all block sizes, because the heuristic proposed in [3] reduces the bipartition modes evaluations when applying high QP-depth values. Consequently, lower QP-depths imply lower bipartition mode skips. The opposite behavior occurs for the RD-list evaluation in DC-only, which is higher for higher QP-depths for all block sizes. It happens because DC-only is planned to encode homogeneous areas efficiently; then, if the encoding block is not homogeneous, the DC-only evaluation is skipped [4]. The RD-list evaluation in TQ also exhibits a similar behavior than the RD-list evaluation in DC-only, where the increment in QP-depth values raises the percentage of time spent in the RD-list evaluation. Lastly, the Intra-picture skip evaluation, the RMD/MPM selection, and the Intra_Contour pattern generation produce little variations in the encoding time for different QP-depths.

This subsection detailed the encoding time of depth map encoding tools in intra-frame prediction and described the most time-consuming steps. The following subsection investigates the scenarios where each encoding tool is selected for being used in an encoding block.

## 4.1.2   Block Sizes and Encoding Mode Distribution

Figure 4.3 shows the block size percentage selected to encode the depth maps according to the QP-depths. This experiment considers results for all videos inside CTC when evaluated under all-intra-frame configuration.

The selection of the block size is highly dependent on the QP-depth value; i.e., high QP-depth values imply selecting bigger blocks (e.g., $32 \times 32$ and $64 \times 64$ block sizes are used more than 75% of times for QP-depth = 45), and the opposite is also true (e.g., $4 \times 4$ and $8 \times 8$ block sizes are used more than 65% of times for QP-depth = 34). Besides, $16 \times 16$ blocks selection percentage remains nearly constant, i.e., independently of the QP-depth. This happens because low QP-depths tend to preserve the depth map details, generating heterogeneous regions, which are encoded better with smaller block sizes. On the other hand, high QP-depths tend to attenuate the depth map details, creating homogeneous regions, which are encoded better with bigger block sizes. Besides, it is crucial to understand the encoder decisions according to the selection of the available encoding flows. This analysis is presented next, considering five possibilities of encoding flows:

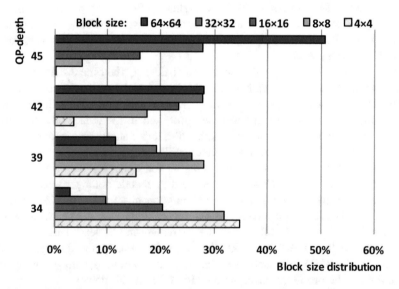

**Fig. 4.3** Block size distribution for four quantization scenarios and five block sizes

- Bp-TQ – a bipartition mode is selected as the final encoder decision using the TQ flow;
- Intra-TQ – the HEVC intra-frame prediction is selected as the final encoder decision using the TQ flow;
- Bp-DC – a bipartition mode is selected as the final encoder decision applying the DC-only flow;
- Intra-DC – the HEVC intra-frame prediction is selected as the final encoder decision applying the DC-only flow, and;
- Intra-picture skip – any Intra-picture skip mode is selected as the final encoder decision.

Figure 4.4 shows that analyzing the two corners of QP-depths in CTC, the mode distribution selection for each block size has a low level of dependency with the QP-depth value.

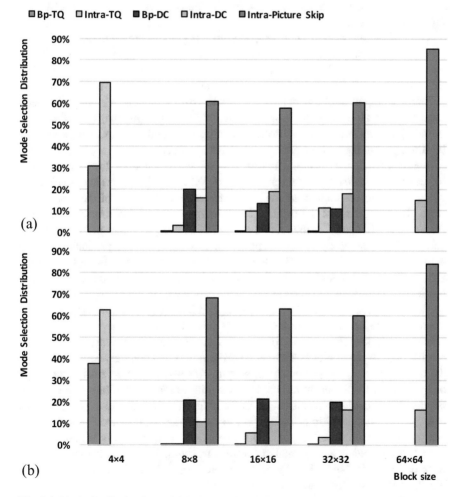

**Fig. 4.4** Mode distribution for each block size with (**a**) QP-depth = 34 and (**b**) QP-depth = 45

Intra- and Bp-TQ flows were selected 68% and 32% of the cases for 4 × 4 block sizes, respectively. This is the unique block size where the Intra- or Bp-TQ flows are used in a relevant number of times since the flows Intra-picture skip and DC-only are not available.

The five encoding flows are evaluated for blocks ranging from 8 × 8 to 32 × 32. In this range, independently of QP-depth, the Intra-picture skip encoding flow was selected more than 50% of times. The DC-only evaluation using Bp-DC and Intra-DC are also significant in this distribution since both encoded flows are selected 33% of times, on average. Intra-TQ and Bp-TQ are the less selected encoding flows. Intra-TQ is selected a maximum of 11% of times (QP-depth = 34 and block size of 32 × 32) and a minimum of less than 1% (QP-depth = 45 and block size of 8 × 8). Bp-TQ was selected less than 1% of times for all scenarios. For these intermediate block sizes, the higher the QP, the higher the use of Intra-picture skip and Bp-DC and the smaller the use of Intra-DC, Intra-TQ, and Bp-TQ. Besides, for these block sizes, Bp-TQ tends to zero for the highest value of QP-depth assessed.

Only Intra-picture skip, Intra-DC, and Intra-TQ modes are evaluated for 64 × 64 block sizes. However, the experimental results demonstrate that the Intra-TQ is almost not used. The Intra-picture skip mode is selected more than 85% of times, followed by Intra-DC. It occurs mainly because larger blocks are chosen when encoding a region with low details, such as the background or the body of objects composed of large areas of homogeneous values in depth maps. In this case, Intra-TQ is almost not used since Intra-picture skip, and Intra-DC can reduce the bitrate expressively without affecting the visual quality.

Figure 4.5a, b, c, d illustrate the selection of the five encoding flows when encoding the central view of the first frame of two CTC videos (Balloons and Shark). The boxes of the figures containing the colors (the same of Fig. 4.4) of the used modes were plotted in front of the original depth image using 50% of transparency. The frames encoded using the lowest and the highest QP-depth values point that higher values of QP-depth decrease the image details significantly, producing several smooth regions. This means that the efficient encoding of depth maps with high QP-depths needs bigger block sizes, as shown in Fig. 4.5. On the other hand, small block sizes are suitable to encode the depth map details using low QP-depths.

Figure 4.5 displays that Bp-TQ, Bp-DC, and Intra-TQ are applied mainly in the borders of the objects, while Intra-picture skip and Intra-DC are selected in the blocks inside smooth areas such as bodies and backgrounds. Intra-DC can also reach a high usage in blocks that can be divided into two homogeneous regions. As the QP-depth rises, the percentage of smooth area also increases, raising the usage of Intra-picture skip and Intra-DC. It is essential to identify the high use of Intra-picture skip encoding mode is concentrated in the background and the bodies of the objects. Intra-picture skip is hardly used in the borders of the objects since it does not transmit residual information. However, Intra-picture skip is suitable on the object borders, when a horizontal or vertical line segments the block. In these cases, Intra-picture skip Mode 0 or Mode 1 are selected, as it can be observed in some blocks of the Balloons video sequence.

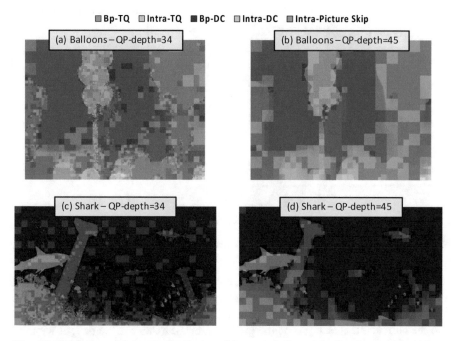

**Fig. 4.5** Illustration of the mode distribution of the central view of the first frame of the videos Balloons (**a**) QP-depth = 34, (**b**) QP-depth = 45, and shark (**c**) QP-depth = 34 and (**d**) QP-depth = 45

## 4.2 Algorithms Designed for the Intra-Frame Prediction of the 3D-HEVC Depth Maps Coding

This section shows the design of four solutions for the encoding time reduction of the intra-frame prediction: (i) DMM-1 Fast Pattern Selection (DFPS) [5]; (ii) Pattern-based Gradient Mode One Filter (P&GMOF) [6]; (iii) Enhanced Depth Rough Mode Decision (ED-RMD) [7]; and (iv) quadtree limitation using decision trees [8].

The design of the DFPS and P&GMOF algorithms focuses on speeding up the Intra_Wedge tool since it is the individual most time-consuming operation. ED-RMD decreases the RD-list size; consequently, reducing the encoding effort of both TQ and DC-only flows. The quadtree limitation was designed utilizing decision trees to avoid unnecessary evaluations of small blocks inside the quadtree.

### 4.2.1 DMM-1 Fast Pattern Selector (DFPS)

Figure 4.6 displays the current encoding PU and the upper and left neighbor PU. A vector called Pattern Vector starts the operation filled with zeros. Always, at the end of the Main Stage of Intra_Wedge, the best pattern number (Pattern$_{curr}$) selected is inserted into the corresponding Pattern Vector position.

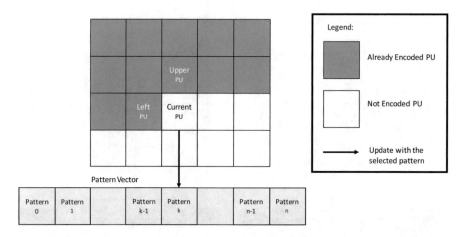

**Fig. 4.6** An example of neighbor PUs and patterns selected in the Intra_Wedge Main Stage

Let Pattern$_{up}$, Pattern$_{left}$ and, Pattern$_{curr}$ be the patterns chosen by the upper, left and current PU in the Main Stage of Intra_Wedge, respectively. Let P$_{copy}$ be the predictor that seeks for the pattern with the smallest distortion between Pattern$_{left}$ and Pattern$_{up}$ to the current PU. Let P$_{extend}$ be the predictor that searches for the pattern with the smallest distortion among extended wedgelet orientations of Pattern$_{left}$ and Pattern$_{up}$. It was performed an experiment associating each block size (i.e., $4 \times 4$, $8 \times 8$, $16 \times 16$ and $32 \times 32$) with a different Pattern Vector aiming to explain the relevance of the predictors P$_{copy}$ and P$_{extend}$, which are defined by Eqs. (4.1 and 4.2).

$$P_{copy}: \quad Pattern_{curr} = PMinDist\left(Pattern_{up}, Pattern_{left}\right) \qquad (4.1)$$

$$P_{extend}: \quad Pattern_{curr} = PMinDist\left(Extend\left(Pattern_{up}, Pattern_{left}\right)\right) \qquad (4.2)$$

Where *PMinDist* searches for the Intra_Wedge Pattern with the smallest distortion and *Extend* returns all patterns with the extended orientation.

The main objective of this experiment is to correlate the decisions of the previous evaluated neighbor PUs with the current evaluation. When evaluating the K$^{th}$ PU in the row, the pattern selected by the Left$_{PU}$ is stored in the position Pattern$_{k-1}$, and the pattern selected by the Upper$_{PU}$ is stored in the position Pattern$_k$ of the Pattern Vector. Thus, when accessing the Pattern Vector, the Current$_{PU}$ can retrieve the required information for the prediction based on P$_{copy}$ and P$_{extend}$. At the end of the Main Stage of the Intra_Wedge, the Current$_{PU}$ writes this information in Pattern$_k$, since the pattern selected by the Upper$_{PU}$ is no longer necessary for the prediction of any other block. Then, when evaluating the next PU (K + 1 in the row), the required information for its prediction are available in the Pattern Vector in the positions Pattern$_k$ and Pattern$_{k+1}$.

Every video sequence available in the CTC was evaluated computing the success rate of each part of each predictor according to the block size. Equation (4.3) describes the success rate, which is the number of cases the proposed predictors had success (i.e., when the predictor obtained the same result of the original encoding flow) divided by the total number of evaluations. In this experiment, the evaluation of $P_{extend}$ predictor is only allowed when $P_{copy}$ fails.

$$\text{Success rate} = \frac{Number\ of\ cases\ the\ predictor\ had\ success}{Total\ number\ of\ evaluations}\ (\%) \tag{4.3}$$

Figure 4.7 presents the statistical results according to the block size. The total success rate is the union of the success rates of all predictors (i.e., it considerers that any of the predictors had success), which is relevant information when applying these predictors together in a single solution. Analyzing these results, one can note that the smaller PUs obtain higher success in $P_{copy}$, while larger PUs get better results in $P_{extend}$. Moreover, the combination of both predictors allows obtaining high success rate independent of the encoding block size.

Along with this analysis, Fig. 4.8 exhibits the probability density function of having success in Pcopy according to the acquired distortion divided by the block size. This metric is used as a criterion since larger block sizes tend to have larger distortion values. This function, which follows a Gaussian distribution, was obtained by executing the GT_Fly video sequence under the All-Intra (AI) frame mode, which was randomly selected to perform this analysis.

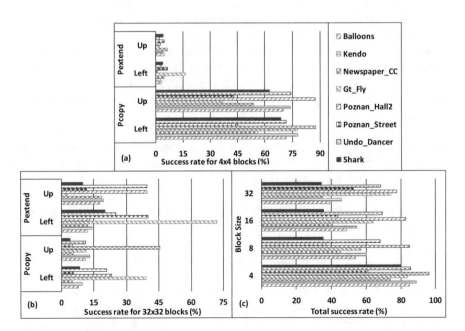

**Fig. 4.7** Success rate for $P_{copy}$ and $P_{extend}$ for (**a**) 4 × 4 and (**b**) 32 × 32 blocks. (**c**) Average total success rate for all available block sizes

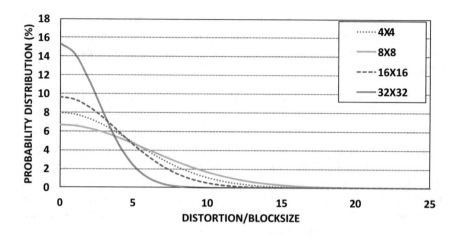

**Fig. 4.8** Probability density function of Pcopy having success, considering the distortion divided by the block size in GT_FLY under AI-frame mode

Figure 4.8 shows a high probability of these predictors in having success for small values of distortion. The probability of success is practically zero for values of distortion divided by the block size larger than 15 since larger distortion values mean the predictors fail and a different pattern encodes the PU, requiring further evaluations.

A similar analysis was done for $P_{extend}$, pointing to similar conclusions. Thus, the values of distortion divided by the block size, achieved by the predictors, are directly related to the predictor success. Consequently, these predictors can be applied to perform an early Intra_Wedge termination decision according to a threshold criterion. Moreover, these experiments enable to gather there is a high probability of the predictors having success by performing at most two wedgelet evaluations in $P_{copy}$ or performing more evaluations in $P_{extend}$, but lower than the traditional encoding flow, for all available block sizes.

Figure 4.9 displays the dataflow model of the designed DFPS algorithms inside the Intra_Wedge encoding algorithm. Looking for a lightweight solution that skips many wedgelet evaluations, DFPS starts getting the minimum distortion in $P_{copy}$ performing at most two pattern evaluations. If the distortion divided by the block size is lower than the THreshold 1 (TH1), which is defined by the offline analysis described later, then the Main Stage of Intra_Wedge is not made, and the refinement is performed followed by the calculation of the residue

Additional evaluations are needed to obtain a reliable prediction on Intra_Wedge Main Stage when TH1 criterion is not met. A medium-weight solution is designed instead of evaluating the entire Intra_Wedge initial set in the Main Stage, where $P_{extend}$ is evaluated extending left and upper PUs wedgelet orientation. Once more, the minimum distortion divided by the block size is compared to THreshold 2 (TH2). If it is smaller than TH2, then the Main Stage is not evaluated, and the Refinement Stage is performed. Otherwise, further evaluations are required, and the remaining wedgelets are evaluated without simplifying the Main Stage.

**Fig. 4.9** DFPS dataflow model

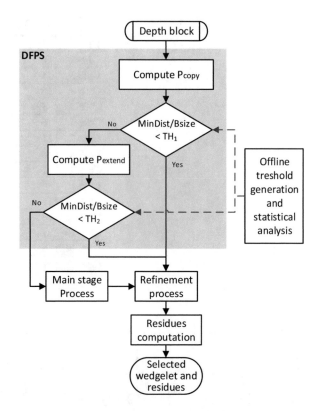

In the worst case (i.e., no simplification is performed), the RD-costs computed by DFPS is the same than the traditional approach. Notice the DFPS algorithm always stores in a vector the information of the pattern selected in the Main Stage. Consequently, even if other modes has been selected in the complete RD-cost evaluation, this information is used to accelerate the encoding of neighbor PUs.

The threshold values affect the encoding timesaving and efficiency significantly. Thus, making extensive experimental analysis of 16 scenarios allows determining the impact of threshold variation using the GT_Fly video sequence again. In each scenario, ten frames of GT_Fly were encoded and compared to the original 3D-HTM results. The average and standard deviation of the distortion divided by the block size has been stored in the previous evaluation of GT_Fly video sequence, and they were used to determine the new evaluation scenarios.

The new evaluation scenarios suppose Eq. (4.4) for each threshold (represented by $TH_n$), where k is selected empirically, ranging from 0 to 3, and $u_n$ and $std_n$ are the average and standard deviation values for the $TH_n$ computation, respectively. The threshold definition and k values were selected for ensuring the predictors have success in 50% (k = 0), 84% (k = 1), 97.5% (k = 2) and, 99.85% (k = 3) of cases. Table 4.2 presents all evaluated thresholds, with $u_1 = 1$, $u_2 = 2$, $std_1 = 5$, and $std_2 = 25$.

$$TH_n = u_n + k \times std_n \tag{4.4}$$

**Table 4.2** Thresholds for the new scenarios

| Case | Threshold 1 (TH₁) | Threshold 2 (TH₂) |
|------|-------------------|-------------------|
| Case-1 | 1 | 2 |
| Case-2 | 6 | 2 |
| Case-3 | 11 | 2 |
| Case-4 | 16 | 2 |
| Case-5 | 1 | 27 |
| Case-6 | 6 | 27 |
| Case-7 | 11 | 27 |
| Case-8 | 16 | 27 |
| Case-9 | 1 | 52 |
| Case-10 | 6 | 52 |
| Case-11 | 11 | 52 |
| Case-12 | 16 | 52 |
| Case-13 | 1 | 77 |
| Case-14 | 6 | 77 |
| Case-15 | 11 | 77 |
| Case-16 | 16 | 77 |

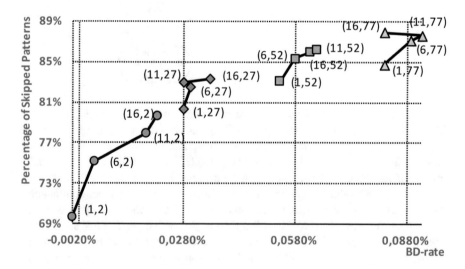

**Fig. 4.10** Percentage of skips on the evaluation of the Intra_Wedge patterns versus the BD-rate impact

Figure 4.10 displays the percentage of wedgelet evaluations skipped versus the BD-rate criterion. DFPS allows some operation points capable of rendering better coding efficiency or higher pattern skips. Case-2 was selected empirically (high-lighted in Fig. 4.10) as the best operation point to be further evaluated. However, other operation points could lead to higher encoding timesaving and higher impact on the video quality.

## 4.2.2 Pattern-Based Gradient Mode One Filter (P&GMOF)

Figure 4.11 exemplifies an 8 × 8 depth block with the gradient values of the block borders and the wedgelet selected when the original Intra_Wedge algorithm is applied. This figure allows understanding the motivation for designing the proposed Pattern-Based Gradient Mode One Filter (P&GMOF) algorithm.

The gradient vectors of the upper (Grad$_{Upper}$), left (Grad$_{Left}$), bottom (Grad$_{Bottom}$) and right (Grad$_{Right}$) rows and columns are obtained by applying the Eqs. (4.5, 4.6, 4.7 and 4.8), respectively; where P(x,y) represents the luminance sample pixel in the position (x, y) of the block. The x value in these equations ranges from 1 to *size*-1, where size represents the width of the block. Notice in the example of Fig. 4.11 that the best encoding wedgelet is found around the positions with the highest gradient. Consequently, one can conclude, the positions of the borders with high gradient values tend to be near the best candidates for Intra_Wedge wedgelet decision process.

$$Grad_{Upper}(x) = |P(1, x) - P(1, x+1)| \qquad (4.5)$$

$$Grad_{Left}(x) = |P(x, 1) - P(x+1, 1)| \qquad (4.6)$$

$$Grad_{Bottom}(x) = |P(size, x) - P(size, x+1)| \qquad (4.7)$$

$$Grad_{Right}(x) = |P(x, size) - P(x+1, size)| \qquad (4.8)$$

**Fig. 4.11** Example of a wedgelet selection and an analysis of its border gradient

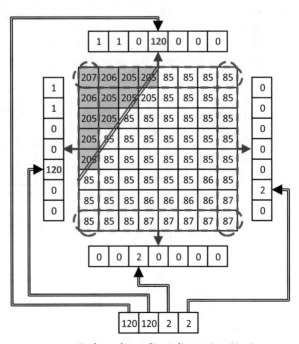

Ordered gradient list using N=4

Since the Intra_Wedge algorithm is one of the most time-consuming operations in depth map intra-frame prediction, the work [9] proposed the Gradient-Based Mode One Filter (GMOF) algorithm to reduce the number of evaluated wedgelets considerably in Intra_Wedge, and some variations of this algorithm are presented in [10]. The algorithms described in [9, 10] reduce the number of wedgelets by searching for wedgelets whose straight line starts in a position with a significant change of the gradient value. The solutions proposed in [9] are line-based because they consider the position of the stored straight line, while the standard execution of Intra_ Wedge performed in 3D-HTM uses the stored binary pattern.

The original line-based Gradient-based Mode One Filter (GMOF) algorithm creates a gradient list with N positions ordered by the highest gradients of the borders. The positions in this list point to the position change that obtained that gradient, as presented at the bottom of Fig. 4.11. Then, the search algorithm evaluates only wedgelets whose straight line is located in those positions. By applying intense experimentation, the best N selected in [9] was 8.

Other GMOF variations are based on the original GMOF algorithm [9]. The Strong GMOF (S-GMOF) algorithm picks the two highest gradients from two arbitrary borders. The Single Degree of Freedom GMOF (SDF-GMOF) selects the highest slope among all possible positions and set a wedgelet line start in it. The ending of the wedgelet is chosen searching in the wedgelet list for the nearest end position compared to the second gradient position. The Double Degree of Freedom GMOF (DDF-GMOF) examines the entire wedgelet list for the wedgelet that has the lowest distance compared to the two highest gradient positions. The S-GMOF, SDF-GMOF, and DDF-GMOF algorithms can reduce the evaluation of the entire wedgelet list to a single wedgelet evaluation. After this initial evaluation, the original Intra_Wedge refinement is applied in all these algorithms. However, often when using these line-based algorithms (including the original GMOF), it is impossible to reach the best wedgelet approximation as exemplified in Fig. 4.12, which shows the encoding a 4 × 4 depth block with the best pattern

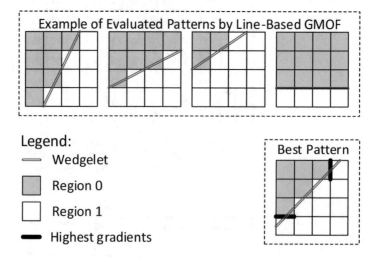

Fig. 4.12  Best and evaluated patterns using Pattern-Based GMOF of a 4 × 4 encoding depth block

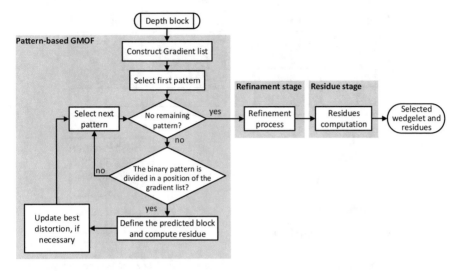

**Fig. 4.13** P&GMOF dataflow model

selected by the original Intra_Wedge algorithm. This example displays the best wedgelet pattern is obtained starting the wedgelet in an intermediary pixel, and there is a gradient value only in a half pixel of distance from there. It occurs because wedgelet evaluations are pattern-based and not line-based as the GMOF algorithm explored.

A new P&GMOF algorithm is designed considering that the Intra_Wedge evaluation does not use the wedgelet line position, but only the binary pattern. Figure 4.13 shows the P&GMOF dataflow model.

The gradient list built by the original line-based GMOF is maintained in this algorithm by applying Eqs. (4.5, 4.6, 4.7 and 4.8). After building this list, the pattern of each wedgelet in the initial wedgelet pattern set is analyzed with the positions in the gradient list. When there is a pattern division in any of those positions (i.e., it changes from one region to another), the wedgelet is evaluated by its RD-cost; otherwise, the next wedgelet is evaluated using the same process. Furthermore, when the initial wedgelet evaluation finishes, the original refinement is applied, and the wedgelet that obtained the lowest RD-cost is selected as the best-encoded wedgelet, sending its residues and encoding information to the next encoder modules.

The same example presented in Fig. 4.12 for line-based GMOF is also presented for P&GMOF in Fig. 4.14. However, the pattern-based version designed here can achieve the same wedgelet that would be selected by the original Intra_Wedge algorithm.

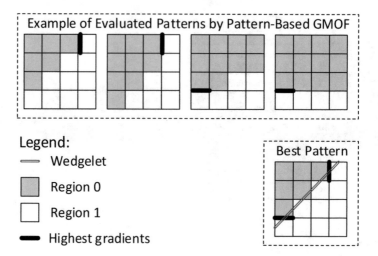

**Fig. 4.14** Best and evaluated patterns using Pattern-Based GMOF of a 4 × 4 encoding depth block

### 4.2.3   Enhanced Depth Rough Mode Decision (ED-RMD)

In the first motivational experiment presented here, all videos inside CTC were encoded using the corners QP-depth in the AI-frame scenario. The best directional mode selected after computing its RD-cost during each block computation was stored and compared to its position in the RD-list for depth map coding. Both, the RMD or MPM algorithms can insert the modes inside RD-list. When inserted by RMD, they are previously ordered by their SATD, as previously described in Sect. 2.2.1. The positioning of these modes in the RD-list is called here as RMD rank.

Figure 4.15a and b present the statistical analysis of each rank in the RMD and MPM selections for QP-depths equal to 34 and 45, respectively. In 4 × 4 and 8 × 8 blocks, the legend "Others" groups the modes ranked in the positions 3–8, respectively. In 16 × 16 to 64 × 64 blocks, the legend "Others" represents only the mode ranked in position three since these block sizes insert only three modes and the MPM inside RD-list as implement in 3D-HTM.

Notice the first RMD is the most representative selection signifying more than 50% in all encoding cases. Moreover, when increasing the encoding QP-depth, the selection of the first RMD mode is still higher. Furthermore, there are other positions in RD-list that contains a significant selection, i.e., the second RMD mode and the first MPM mode.

The statistical analyses demonstrate the depth map coding does not require the complete RD evaluation for all modes inserted into RD-list because the RMD and MPM pre-analysis were designed considering only texture coding, which contains a more complex behavior. Besides, due to many new modes inserted into the intra-frame prediction of depth maps, the removal of some modes in the RD evaluation of the HEVC intra-frame prediction leads to significant encoding timesaving results without considerable impact on the encoding efficiency.

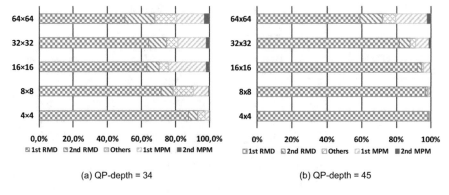

(a) QP-depth = 34          (b) QP-depth = 45

**Fig. 4.15** Statistical analysis of RMD and MPM selections

Forty-eight cases running 3D-HTM software were analyzed showing that several encoding modes selected by the RMD and MPM algorithms have almost no impact in the encoding selection. In each execution case, the number of modes selected by RMD and MPM was varied to identify the encoding efficiency and timesaving of the proposed ED-RMD heuristic. These experiments encoded ten frames of Balloons and Undo_Dancer sequences under the AI configuration. In these experiments, the number of modes evaluated in RD-list was varied between one and eight for $4 \times 4$ and $8 \times 8$ blocks, and between one and three for higher block sizes ($16 \times 16 - 64 \times 64$). Besides, the experiments analyze the use of one and two MPMs.

Tables 4.3 and 4.4 describe the results of the experiments using one and two MPMs, respectively. The "Encoding time" column represents the percentage of encoding time compared to the original 3D-HTM execution, considering the entire encoder time (i.e., both texture and depth map encoding time). Notice that removing one intra-frame mode from larger block sizes implies in higher reduction on the computational effort than removing one mode from smaller block sizes. Moreover, removing one evaluation of MPM implies in a small computational reduction and a negligible impact on BD-rate. Consequently, when aiming to design real-time systems, reducing the MPM evaluation to a single mode results in a sound tradeoff.

These experiments show that different system configuration can lead to different computational effort reduction. Notice there is no optimum configuration for the proposed algorithm, and the system can be configured according to its application constraints such as maximum BD-rate, encoding time, or maximum power dissipation. If a higher encoding efficiency is required, then the insertion of more modes into the RD-list is necessary; however, when requirements of decreasing the encoding time or power dissipation are critical, fewer modes can be inserted in the RD-list with small impact on the encoding efficiency.

As a decision from the previous experiment, the ED-RMD heuristic was defined using the four best-ranked modes in RMD and one in MPM for $4 \times 4$ and $8 \times 8$ blocks, and only the best-ranked mode in RMD and one in MPM for $16 \times 16 - 64 \times 64$ blocks, regarding the analysis described before. Some other operation points could also be explored; however, this configuration obtained the best tradeoff

**Table 4.3** ED-RMD experiment allowing the evaluation of one MPM

| Mode selected for 4 × 4 and 8 × 8 | Mode selected for 16 × 16 to 64 × 64 | | | | | |
|---|---|---|---|---|---|---|
| | 1 | | 2 | | 3 | |
| | BD-rate | Encoding time | BD-rate | Encoding time | BD-rate | Encoding time |
| 8 | 0.164% | 90.2% | 0.072% | 95.5% | 0.043% | 99.3% |
| 7 | 0.146% | 87.8% | 0.120% | 92.7% | 0.052% | 97.0% |
| 6 | 0.219% | 85.1% | 0.159% | 90.4% | 0.097% | 94.5% |
| 5 | 0.223% | 82.6% | 0.143% | 87.7% | 0.109% | 91.9% |
| 4 | 0.290% | 80.2% | 0.206% | 85.3% | 0.175% | 89.5% |
| 3 | 0.380% | 77.6% | 0.263% | 82.7% | 0.239% | 87.0% |
| 2 | 0.440% | 74.9% | 0.407% | 80.1% | 0.326% | 84.2% |
| 1 | 0.577% | 71.4% | 0.455% | 76.4% | 0.451% | 80.5% |

**Table 4.4** ED-RMD experiment allowing the evaluation of two MPM

| Mode selected for 4 × 4 and 8 × 8 | Mode selected for 16 × 16 to 64 × 64 | | | | | |
|---|---|---|---|---|---|---|
| | 1 | | 2 | | 3 | |
| | BD-rate | Encoding time | BD-rate | Encoding time | BD-rate | Encoding time |
| 8 | 0.108% | 90.7% | 0.050% | 95.8% | 0.000% | 100.0% |
| 7 | 0.103% | 88.3% | 0.053% | 93.5% | −0.002% | 97.7% |
| 6 | 0.149% | 85.9% | 0.136% | 91.0% | 0.020% | 95.3% |
| 5 | 0.176% | 83.3% | 0.128% | 88.4% | 0.068% | 92.6% |
| 4 | 0.248% | 80.8% | 0.194% | 86.1% | 0.120% | 90.2% |
| 3 | 0.365% | 78.4% | 0.274% | 83.5% | 0.235% | 87.7% |
| 2 | 0.432% | 75.5% | 0.347% | 80.7% | 0.280% | 85.0% |
| 1 | 0.558% | 71.7% | 0.478% | 77.1% | 0.397% | 81.3% |

between computational effort decrease and coding efficiency degradation. The selected configuration saved almost 20% of the time, with a BD-rate increase under 0.3%, which is a significant computational effort reduction and a negligible influence on the encoding efficiency.

### 4.2.4   Applying Machine Learning for Quadtree Limitation

Many encoder attributes were investigated aiming to determine the most relevant ones to build the static Coding Unit (CU) trees for encoding depth maps. A significant amount of data from the video sequences of depth maps and internal encoding variables were collected to find features that could identify the splitting decisions. The parameters listed in Table 4.5 were evaluated and stored for each CU size during the 3D-HTM encoder execution.

**Table 4.5** Parameters evaluated in I-frames and their description

| Parameter | Description |
|---|---|
| QP-depth | The current QP-depth value, which is responsible for defining the compression rate and has much impact on the CU split decision |
| RD-cost | The RD-cost obtained when encoding the current CU size, which was used to evaluate the relations among the different encoder decisions and the encoding efficiency |
| VAR | The variance of the original samples of the current CU, which indicates the block homogeneity, and then, if the block should or not be split |
| VAR_size | The maximum variance of smaller blocks inside of the current CU, which represents the maximum variance of the samples inside a block. For a $64 \times 64$ CU, there are four instances of this attribute, one for each possible block size ($4 \times 4$, $8 \times 8$, $16 \times 16$ and $32 \times 32$). This information indicates the homogeneity or presence of edges into smaller blocks |
| Average | The average value of the current CU samples. This information indicates if the encoding CU is near or far from the camera. Additional details of near objects should be maintained and, in this case, it is interesting to evaluate lower CUs sizes |
| MaxDiff | The highest maximum difference between all samples of the current CU, which is useful in the CU split decisions since this information indicates sudden variations in samples values |
| Corners_grad | The maximum absolute difference of the four corners in the current CU. This information indicates the presence of edges in the current CU |
| Max_Grad | Maximum gradient is the maximum absolute difference of the four corners of smaller blocks inside of the current CU. Similar to VAR_size, there are four instances of this attribute for $64 \times 64$ CU, and they indicate when a CU should be split or not |

These attributes indicate depth map blocks that tend to be harder to encode and, consequently, tend to produce a splitting decision. Figure 4.16 displays the density probability of the $64 \times 64$ CUs do not be split into smaller CUs for some collected attributes. Figure 4.16a and b show MaxDiff and $VAR_{64}$ have lower values for those CUs that are not split into smaller CUs. The probability of RD-cost is shown in Fig. 4.16c and indicates a high similarity with the splitting decision. Figure 4.16d shows the distribution of $VAR_{16}$, which provides essential information for sub-blocks inside a $64 \times 64$ CU since high values of $VAR_{16}$ can indicate the presence of edges in the current CU and its information can be hidden when assessing the variance of larger blocks. In the case of sub-blocks with low-variance values, the current encoding CU tends not to be split into smaller CUs.

The attribute evaluation permits concluding only QP-depth, RD-cost, VAR, VAR_size, Average, and MaxDiff are essential to building the static CU decision trees. Consequently, the process of data mining training only employed these attributes.

Since the 3D-HEVC intra-prediction of depth maps allows encoding only square CU sizes from $8 \times 8$ up to $64 \times 64$, three static decision trees were designed defining when CUs of sizes $16 \times 16$, $32 \times 32$ and $64 \times 64$ should be or not split into smaller CUs.

The Kendo video sequence was encoded in AI configuration, considering all CTC QP values, for the data mining process. The Coding Tree Unit (CTU) size has been limited to $16 \times 16$, $32 \times 32$, and $64 \times 64$ pixels for each evaluation. The following

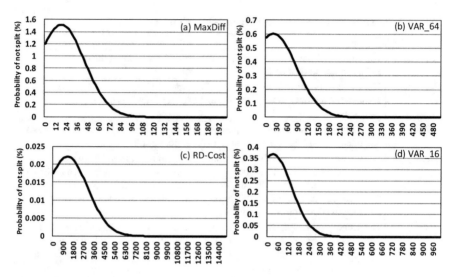

**Fig. 4.16** Probability density of the analyzed attributes does not divide the current 64 × 64 CUs

data was stored for each encoded CU: (i) all information presented earlier in Table 4.5, and (ii) the information indicating if the CU has been split or not. The Kendo video sequence was randomly selected from the CTC dataset, and it was used to extract the data necessary to the offline training process. Although only one sequence was used in this training, we repeated our methodology using more and different video sequences, and we obtained similar results; however, these additional evaluations were omitted in this book.

We used the Waikato Environment for Knowledge Analysis (WEKA) [11], version 3.8, to train each decision tree with the J48, which is an open-source implementation of the C4.5 algorithm [12] available on the WEKA software. Looking for better data balancing, the input files were organized in two data sets with equal sizes having inputs that result in splitting and not splitting the CUs. Besides, to avoid the overfitting problem on the train data set, the Reduced Error Pruning (REP) [13] was performed in each decision tree, reducing the depth of the trees and allowing a better generalization.

Figure 4.17 exemplifies the static decision tree obtained for 64 × 64 CUs, where the leaves "N" and "S" are not split and split decisions, respectively. The decision trees for 32 × 32 and 16 × 16 CUs encompass five and eight decision levels, respectively. The attributes in each decision tree were selected through the information gain, which is used by the training algorithm of the WEKA decision trees [11].

Table 4.6 indicates the entire list of attributes and the corresponding usage in the three proposed decision trees. These attributes are selected in the decision trees building that uses the Information Gain (IG). The IG of each attribute refers to the difference between the order of entropy for the entire data set and the entropy of the partitioned subset for the attribute being evaluated. Table 4.6 also presents the IG of each attribute.

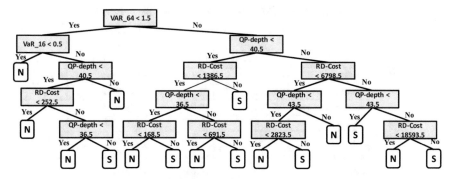

**Fig. 4.17** Decision tree for splitting decision in 64 × 64 CUs

**Table 4.6** Attributes used in the decision trees for I-frames

| Attributes | 64 × 64 | | 32 × 32 | | 16 × 16 | |
|---|---|---|---|---|---|---|
| | Used | IG | Used | IG | Used | IG |
| QP-depth | × | 0.089 | × | 0.123 | × | 0.039 |
| RD-cost | × | 0.312 | × | 0.203 | × | 0.293 |
| Var | × | 0.395 | × | 0.17 | × | 0.027 |
| Var_16 | × | 0.397 | | – | | – |
| Var_8 | | – | × | 0.171 | × | 0.025 |
| Var_4 | | – | | – | × | 0.025 |
| MaxDiff | | – | | – | × | 0.026 |
| Average | | – | × | 0.11 | × | 0.061 |
| Accuracy | 92.68% | | 83.59% | | 84.54% | |

The proposed solution does not raise the computational effort to the 3D-HEVC encoder since the training of the static trees is an offline operation, which is performed only once. The accuracies obtained by each decision tree are presented in Table 4.6, which are 84.54%, 83.59%, and 92.68% for 16 × 16, 32 × 32, and 64 × 64 CUs, respectively.

## 4.3  Experimental Results

Table 4.7 illustrates the experimental results of AI-frame evaluation for DFPS Case-2 and P&GMOF, and Table 4.8 presents the results for ED-RMD and quadtree limitation using machine-learning algorithms. Both results consider the CTC described in Sect. 2.5. One can notice there is some significant performance variation in different videos execution for all evaluated algorithms, which is explained because each video has specific characteristics, as highlighted in Sect. 2.5.

In average, the DFPS solution reduces 11.7% the encoder time considering the texture and depth coding execution in AI-frame configuration, with a drawback of

**Table 4.7** DFPS and Pattern-based GMOF evaluation in the AI-frame scenario

| Resolution | Video | DFPS Case-2 Synthesis only BD-rate | Timesaving | P&GMOF Synthesis only BD-rate | Timesaving |
|---|---|---|---|---|---|
| 1024 × 768 | Balloons | 0.0152% | 8.1% | 0.0158% | 6.8% |
| | Kendo | 0.0189% | 6.5% | 0.0141% | 5.1% |
| | Newspaper_cc | 0.0733% | 10.8% | 0.0344% | 6.6% |
| | Average | 0.0358% | 8.5% | 0.0214% | 6.2% |
| 1920 × 1088 | GT_Fly | 0.0045% | 15.3% | 0.0049% | 8.3% |
| | Poznan_Hall2 | 0.1847% | 13.3% | −0.0018% | 6.1% |
| | Poznan_Street | 0.0380% | 14.6% | 0.0051% | 7.2% |
| | Undo_Dancer | 0.0298% | 13.7% | 0.0107% | 5.7% |
| | Shark | 0.0152% | 11.0% | 0.0049% | 8.3% |
| | Average | 0.0533% | 13.6% | 0.0047% | 6.8% |
| Average | | 0.0467% | 11.7% | 0.0119% | 6.5% |

**Table 4.8** ED-RMD and quadtree limitation using machine learning in the AI-frame scenario

| Resolution | Video | ED-RMD Synthesis only BD-rate | Timesaving | Quadtree limitation using machine learning Synthesis only BD-rate | Timesaving |
|---|---|---|---|---|---|
| 1024 × 768 | Balloons | 0.2662% | 19.9% | 0.1425% | 45.7% |
| | Kendo | 0.2844% | 19.4% | 0.1909% | 49.8% |
| | Newspaper_cc | 0.3064% | 20.3% | 0.1062% | 37.3% |
| | Average | 0.2857% | 19.9% | 0.1465% | 44.3% |
| 1920 × 1088 | Gt_fly | 0.0965% | 21.0% | 0.0561% | 58.8% |
| | Poznan_Hall2 | 0.2046% | 22.0% | 0.6242% | 63.4% |
| | Poznan_Street | 0.1484% | 22.7% | 0.1204% | 55.8% |
| | Undo_Dancer | 0.0974% | 19.0% | 0.1406% | 56.5% |
| | Shark | 0.0829% | 20.3% | 0.0352% | 51.9% |
| | Average | 0.1367% | 21.0% | 0.2353% | 57.3% |
| Average | | 0.1671% | 20.6% | 0.1770% | 52.4% |

0.0467% in BD-rate. The P&GMOF was evaluated using N = 8 since this value was already evaluated in previous work [9]. This algorithm saves 6.5% of the encoding time, considering the coding execution time of texture and depth maps, with a BD-rate increase of 0.0119%. Notice the DFPS and P&GMOF algorithms, which are focused on Intra_Wedge, can be applied together to achieve higher timesaving results. In this case, instead of evaluating the remaining patterns when the $P_{extend}$ predictor fails, it is possible to evaluate only the positions indicated by the Pattern-based GMOF.

Figure 4.18 shows the percentage of wedgelets skipped for P&GMOF and DFPS, according to the evaluated video sequence. On average, P&GMOF reduces the wedgelet evaluation by 58%, while DFPS provides a wedgelet evaluation reduction of

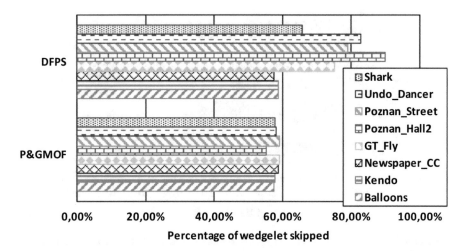

**Fig. 4.18**  P&GMOF and DFPS solutions for reducing the wedgelet evaluation

71%. In the DFPS, the highest wedgelet evaluation reductions are achieved in higher resolution videos because the first predictor succeeds more often in this kind of videos. Although P&GMOF and DFPS obtain a similar average reduction in wedgelet evaluations for lower resolution videos, the superior timesaving of DFPS is explained because it reduces more than the traditional GMOF the effort spent in smaller block sizes.

ED-RMD was configured with the highest encoding timesaving obtained in the analysis with a BD-rate under 0.3% as described in Sect. 4.2.3. The selected condition uses the four best-ranked modes in the RMD selection and one in the MPM for $4 \times 4$ and $8 \times 8$ blocks. When encoding $16 \times 16 - 64 \times 64$ blocks, the proposed algorithm uses only the best-ranked mode in the RMD selection and the first MPM since this configuration reduces almost 20% the encoding time in the experiments described in Sect. 4.2.3. The previous evaluation of ED-RMD shown in Sect. 4.2.3 was executed only using ten frames of two videos. Then, here all videos and frames were evaluated in CTC AI-frame configuration. On average, it reduced 20.6% of the encoder execution time considering texture and depth execution time with a drawback of 0.1671% in the BD-rate of the synthesized views.

The quadtree limitation with machine learning reduces 52.4% of the encoding time considering the execution of texture and depth maps, on average, with 0.1770% of BD-rate increase. If only the execution time of the depth maps is considered, then this solution achieves 59% of execution reduction. Figure 4.19 depicts the percentage of CUs that were not split according to the CU size and QP-depth value. Higher QPs achieve higher not splitting percentages. For example, concerning $64 \times 64$ CUs and QP = 45, 85% of CUs were not split; it occurs because this QP generates higher compression rates and tends to encode the CTUs with larger CU sizes. The application of decision trees obtained the highest reduction in the encoding time among the designed solutions because it works in a high-level, pruning some levels of quadtree

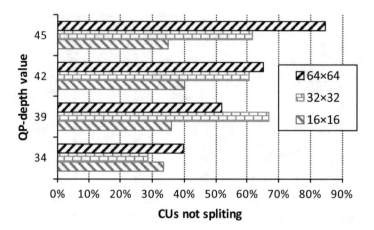

**Fig. 4.19** Percentage of non-splitting CUs according to the block size and QP-depth for AI configuration

evaluations. However, all proposed solutions can be employed in different scenarios. For example, DFPS and P&GMOF can be applied in the design of an Intra_Wedge architecture focusing on decreasing the hardware requirements such as low area consumption, high-frequency operation, and low power dissipation.

Table 4.9 compares the proposed algorithms with the related works mentioned in Sect. 3.1. Some solutions describe results for Random Access (RA) evaluation only, and some other only informs the coding execution time of the depth maps (without considering the entire encoder time).

The work [14] does not provide results for the intra-frame prediction algorithm individually; thus, it is compared to Chap. 5 solutions that explore timesaving algorithms of the inter-frame prediction.

All designed solutions present a small impact in the encoding performance; solutions such as DFPS and P&GMOF provide a significant encoding time reduction with a negligible BD-rate increase (less than 0.05%). Comparing these solutions to other works that focus only on Intra_Wedge, such as [9, 10, 15], the timesaving results provided by the solutions proposed here show a worthy tradeoff reducing Intra-Wedge encoding effort.

The ED-RMD and the quadtree limitation also presents competitive results regarding BD-rate, where an increment smaller than 0.18% is required with a significant reduction in the encoding time, mainly in the quadtree limitation solution. Typically, solutions that provide higher encoding time savings, such as ED-RMD and the quadtree limitation, require a significant BD-rate increase, such as [16–19].

Analyzing the quadtree limitation individually with other solutions that proposed to limit the quadtree, such as [16, 17], the proposed solution obtained a smaller BD-rate increase and provided a higher timesaving result. Compared to [20], the proposed quadtree solution obtained a higher BD-rate increase; however, it is possible reducing significantly more the encoding timesaving, since their results consider only the encoding time of depth maps.

**Table 4.9** Comparisons of the timesaving algorithms of the intra-frame prediction

| Work | 3D-HTM version | Decision level and focused tools | Synthesis only BD-rate | Encoding timesaving |
|---|---|---|---|---|
| DFPS | 16.0 | Mode | 0.047% | 11.7% |
| P&GMOF | | Mode | 0.012% | 6.5% |
| ED-RMD | | Mode | 0.167% | 20.6% |
| Quadtree limitation using machine learning | | Quadtree | 0.177% | 52.4% |
| H. Chen et al. [20] | 16.1 | Quadtree | 0.09% | 45.1% |
| R. Conceição et al. [14] | 16.0 – RA | Block | 0.062% ~0.409% | 9.6% ~ 13.8% |
| C.-H. Fu et al. [15] | 8.1 | Mode | 0.49% | 55% (Intra_Wedge only) |
| Z. Gu et al. [3] | – | Block | Default | Default |
| Z. Gu et al. [4] | – | Mode | Default | Default |
| K.-K. Peng et al. [16] | 13.0 | Mode, Block and Quadtree | 0.8% | 37.6% |
| M. Saldanha et al. [10] | 10.2 – RA | Mode | 0.33% ~1.47% | 4.9% ~ 8.2% (depth only) |
| G. Sanchez et al. [9] | 7.0 – RA | Mode | −0.047% | 9.8% (depth only) |
| G. Sanchez et al. [21] | 7.0 | Block | −0.064% | 5.9% |
| H.-B. Zhang et al. [17] | 13.0 | Quadtree | 0.44% | 41% |
| H.-B. Zhang et al. [18] | 8.1 | Mode | 1.03% | 27.9% (depth only) |
| H.-B. Zhang et al. [19] | 13.0 | Mode and block | 0.54% | 32.9% (depth only) |

# References

1. Sanchez, G., R. Cataldo, R. Fernandes, L. Agostini, and C. Marcon. 2016. 3D-HEVC depth maps intra prediction complexity analysis. *IEEE International Conference on Electronics, Circuits and Systems*: 348–351.
2. Sanchez, G., J. Silveira, L. Agostini, and C. Marcon. 2018. Performance analysis of depth intra coding in 3D-HEVC. *IEEE Transactions on Circuits and Systems for Video Technology* 1–1: 1–12.
3. Gu, Z., J. Zheng, N. Ling, and P. Zhang. 2013. Fast intra prediction mode selection for intra depth map coding, Technical Report, ISO/IEC JTC1/SC29/WG11, 4.
4. ———. 2015. Fast segment-wise DC coding for 3D video compression. *IEEE International Symposium on Circuits and Systems*: 2780–2783.
5. Sanchez, G., L. Jordani, C. Marcon, and L. Agostini. 2016. DFPS: A fast pattern selector for depth modeling mode 1 in three-dimensional high-efficiency video coding standard. *Journal of Electronic Imaging* 25 (6): 063011.
6. Sanchez, G., L. Agostini, and C. Marcon. 2018. A reduced computational effort modelevel scheme for 3D-HEVC depth maps intra-frame prediction. *Journal of Visual Communication and Image Representation* 54 (1): 193–203.
7. ———. 2017. Complexity reduction by modes reduction in RD-list for intra-frame prediction in 3D-HEVC depth maps. *IEEE International Symposium on Circuits and Systems*: 1–4.
8. Saldanha, M., G. Sanchez, C. Marcon, and L. Agostini. 2018. Fast 3D-Hevc depth maps intraframe prediction using data mining. *IEEE International Conference on Acoustics, Speech and Signal Processing*: 1738–1742.

9. Sanchez, G., M. Saldanha, G. Balota, B. Zatt, M. Porto, and L. Agostini. 2014. A complexity reduction algorithm for depth maps intra prediction on the 3D-HEVC. *IEEE Visual Communications and Image Processing Conference*: 137–140.
10. Saldanha, M, B. Zatt, M. Porto, L. Agostini, G. Sanchez. 2016. Solutions for DMM-1 complexity reduction in 3D-HEVC based on gradient calculation. In: IEEE 7th Latin American Symposium on Circuits & Systems, 211–214.
11. Hall, M., E. Frank, G. Holmes, B. Pfahringer, P. Reutemann, and I.H. Witten. 2009. The WEKA data mining software: An update. *ACM SIGKDD Explorations Newsletter* 11 (1): 10–18.
12. Quinlan, J.R. 2014. *C4. 5: programs for machine learning*, 302. Elsevier.
13. Brunk, C.A., and M.J. Pazzani. 1991. An investigation of noise-tolerant relational concept learning algorithms. *Machine Learning Proceedings*: 389–393.
14. Conceição, R., G. Avila, G. Corrêa, M. Porto, B. Zatt, and L. Agostini. 2016. Complexity reduction for 3D-HEVC depth map coding based on early skip and early DIS scheme. *IEEE International Conference on Image Processing*: 1116–1120.
15. Fu, C.-H., H.-B. Zhang, W.-M. Su, S.-H. Tsang, and Y.-L. Chan. 2015. Fast wedgelet pattern decision for DMM in 3D-HEVC. *IEEE International Conference on Digital Signal Processing*: 477–481.
16. Peng, K.-K., J.-C. Chiang, and W.-N. Lie. 2016. Low complexity depth intra coding combining fast intra mode and fast CU size decision in 3D-HEVC. *IEEE International Conference on Image Processing*: 1126–1130.
17. Zhang, H.-B., Y.-L. Chan, C.-H. Fu, S.-H. Tsang, and W.-C. Siu. 2016. Quadtree decision for depth intra coding in 3D-HEVC by good feature. *IEEE International Conference on Acoustics, Speech and Signal Processing*: 1481–1485.
18. Zhang, H.-B., C.-H. Fu, Y.-L. Chan, S.-H. Tsang, and W.-C. Siu. 2015. Efficient depth intra mode decision by reference pixels classification in 3D-HEVC. *IEEE International Conference on Image Processing*: 961–965.
19. Zhang, H.-B., S.-H. Tsang, Y.-L. Chan, C.-H. Fu, and W.-M. Su. 2015. Early determination of intra mode and segment-wise DC coding for depth map based on hierarchical coding structure in 3D-HEVC. *Asia-Pacific Signal and Information Processing Association Annual Summit and Conference*: 374–378.
20. Chen, H., C.-H. Fu, Y.-L. Chan, and X. Zhu. 2018. Early intra block partition decision for depth maps in 3D-HEVC. *IEEE International Conference on Image Processing*: 1777–1781.
21. Sanchez, G., M. Saldanha, G. Balota, B. Zatt, M. Porto, and L. Agostini. 2014. Complexity reduction for 3D-HEVC depth maps intra-frame prediction using simplified edge detector algorithm. *IEEE International Conference on Image Processing*: 3209–3213.

# Chapter 5
# Novel Encoding Time Reduction Algorithms for the 3D-HEVC Inter-Frame Prediction

This chapter presents the book contributions related to the encoding time reduction of the inter-frame prediction. We designed three algorithms in this context: (i) Edge-Aware Depth Motion Estimation (E-ADME) (Sect. 5.1) [1], (ii) an early termination block-level decision (Sect. 5.2) [2], and (iii) a quadtree limitation (Sect. 5.3) [3].

## 5.1 Edge-Aware Depth Motion Estimation (E-ADME)

Figure 5.1a and b displays a slice of two consecutive depth maps of Shark video sequence (frames 2 and 3 of the view 9). Figure 5.1b shows that the Full Search (FS) algorithm was applied in two detached blocks with 16 × 16 pixels, whose search area of [−40, 41] are displayed in Fig. 5.1a with the same color. The yellow box in Fig. 5.1b starts in the pixel (240, 340) and the red box starts in the pixel (240, 440). Figure 5.1c and d show the heat map of the Sum of Absolute Differences (SAD) for the Motion Estimation (ME) search process of a given encoding block in homogeneous (red box in Fig. 5.1b) and edge regions (yellow box in Fig. 5.1b), respectively. The lowest SAD values are denoted by dark blue regions, whereas red regions represent the higher SAD values, meaning the best and worst values, respectively.

The heat map has smooth changes when encoding a homogeneous regions block, with large regions containing low SAD values around the center of the search area. Therefore, homogeneous regions in the depth maps can be encoded using center-biased (i.e., starts searching in the center of the exploration area) lightweight fast ME algorithms. Examples of these algorithms are the Iterative Small Diamond Search Pattern (I-SDSP) or Diamond Search (DS) [4], as described in the works [5, 6], which achieve near-optimal results with less SAD comparisons than the Test Zone Search (TZS) algorithm. However, the SAD analysis of the edge heat map (Fig. 5.1d) shows higher pattern variability, requiring more comparisons and sophisticated ME algorithms for providing higher quality on the synthesized views.

© Springer Nature Switzerland AG 2020
G. Sanchez et al., *Algorithms for Efficient and Fast 3D-HEVC Depth Map Encoding*, https://doi.org/10.1007/978-3-030-25927-3_5

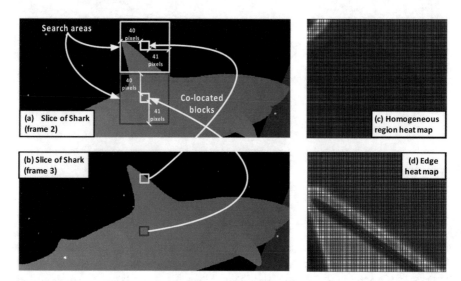

**Fig. 5.1** Slices of the Shark video with (**a**) two detached search areas and (**b**) two detached encoding blocks. SAD heat map for (**c**) homogeneous and (**d**) edge regions

Therefore, designing a scheme able to classify if the encoding block as an edge or homogeneous can reduce the computational effort of the ME and also preserve the encoding efficiency since it considers the encoding block characteristic. This scheme can perform a more sophisticated algorithm for edge regions or a simple algorithm for homogeneous regions.

### 5.1.1    Designed E-ADME Scheme

Figure 5.2 illustrates the E-ADME scheme, which starts using the Simplified Edge Detector (SED) algorithm to identify if the encoding block contains an edge or a homogeneous region. When the encoding block is classified as homogeneous, the lightweight I-SDSP algorithm is applied due to its efficiency for this kind of scenario. I-SDSP can search for low SAD values around the center-biased position; thus, speeding up the ME encoding flow. When SED classifies the encoding block as an edge, the original encoding flow is performed using the TZS algorithm, because it raises the probability of obtaining lower SAD values distant from the co-located block.

The work [5] shows using I-SDSP instead of TZS in ME of depth maps is capable of obtaining a good tradeoff between computational effort and encoding efficiency. However, the ME process of depth map blocks in edge regions is harder than for homogeneous regions because many candidate blocks with low SAD values are found distant from the center-biased block and, thus, I-SDSP can obtain an inefficient prediction. Considering these two aspects, the usage of TZS in edge blocks tends to

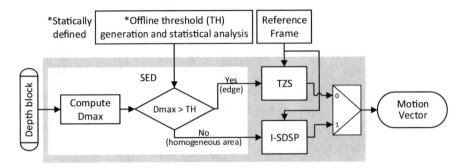

**Fig. 5.2** Scheme proposed for reducing the E-ADME depth map complexity

obtain a sound tradeoff regarding encoding quality and computational effort. One can notice the presented solution does not increase the computational effort or memory accesses compared to the inter-frame prediction of the 3D-HEVC Test Model (3D-HTM) depth maps.

The SED algorithm used as a base in this scheme was described earlier in this book in Chap. 3. The efficiency of the SED algorithm is dependent on the threshold defined statically according to the block size and the video resolution [7]. In [7], the thresholds (TH) for classifying a block as homogeneous or edges were determined according to the block size. These thresholds were interpolated, generating Eqs. (5.1) and (5.2) that shows the polynomials for 1024 × 768 and 1920 × 1088 video resolutions, respectively; W represents the block width and TH the resulting threshold.

$$TH = ceil\left(-0.0186 \times W^2 + 2.2 \times W + 3.5\right) \tag{5.1}$$

$$TH = ceil\left(-0.0038 \times W^2 + 0.74 \times W + 5.1\right) \tag{5.2}$$

In our previous work [7], it was used only four blocks sizes: 4 × 4, 8 × 8, 16 × 16 and 32 × 32, while ME also requires encoding 64 × 64 blocks, asymmetrical blocks such as 64 × 32, 32 × 24, 16 × 12, and others. Therefore, it is required to define new thresholds for those block sizes. The thresholds for the new block sizes were generated using the polynomial interpolation of Lagrange to make a second-degree polynomial equation, which computes the new thresholds based on the fixed thresholds described in [7] and the correspondent block width. The new thresholds were generated for blocks with the widths of 12, 24, and 64 samples.

## 5.2 Early Termination Block-Level Decision Algorithm

Figure 5.3 presents experimental results for analyzing the 3D-HEVC encoder behavior in inter-frame prediction. These experiments were performed following the Common Test Conditions (CTC) (described in Sect. 2.5) considering Random Access (RA) encoder configuration. Our first experiment evaluation allows

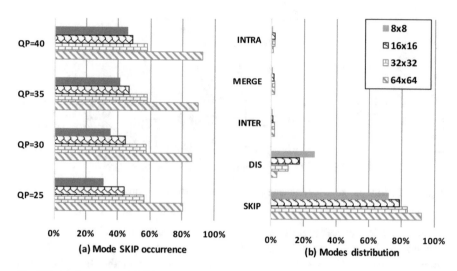

**Fig. 5.3** (**a**) Occurrence of Skip mode for texture encoding and (**b**) modes distribution for depth map encoding when $M_{text}$ = Skip

understanding the Skip mode occurrence in texture encoding considering different QPs. Figure 5.3a shows the Skip mode has a high probability of occurrence in texture encoding for all cases evaluated, ranging from 27% (8 × 8 CUs and QP = 40) to 95% (64 × 64 Coding Units (CUs) and QP = 40). Besides, for all evaluated QPs, more than 90% of the 64 × 64 CUs were encoded using the Skip mode.

In our second experiment, the depth map coding behavior is verified when the correlated reference block in texture view is encoded with the Skip mode. For convenience, let $M_{text}$ be the selected mode by the correlated reference block in texture view. Figure 5.3b displays the mode distribution for depth map encoding when $M_{text}$ = Skip, showing a low probability, less than 4% (on average), of the modes INTER, MERGE or, INTRA being chosen for encoding a current depth map CU. On the other hand, 93% of 64 × 64 CUs and more than 80% (on average) of depth maps CUs are encoded using Skip mode. Besides, more than 15% (on average) select the Intra-picture skip mode for encoding the depth map CUs. Therefore, when $M_{text}$ = Skip, the probability of the encoding depth map CU be Skip or Intra-picture skip is more than 95%, on average.

Figure 5.4 shows the Probability Density Function (PDF) of Skip or Intra-picture skip mode (Skip/Intra-picture skip) being selected as the best encoding mode, when $M_{text}$ = Skip, according to the RD-cost divided by CU width size obtained by evaluating only these two modes. The division of RD-cost by the CU width size has been applied as an evaluation criterion since larger CU sizes tend to have larger RD-cost values to compute. The analysis presented in Fig. 5.4 considers 64 × 64 CUs with the Undo_Dancer video sequence encoding under RA configuration with QP = 30/39 ($QP_{texture}/QP_{depth}$). Similar results were obtained for the other CU sizes and QPs.

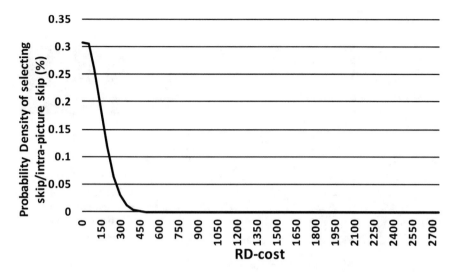

**Fig. 5.4** PDF of encoding depth map CU with Skip or Intra-picture skip according to the RD-cost divided by the CU width size

These results show a high probability of the current CU being encoded using Skip or Intra-picture skip mode for small values of RD-cost when $M_{text}$ = Skip. Moreover, for higher values of RD-cost, Skip and Intra-picture skip modes have no chance to be selected, since higher values of RD-cost mean another encoding mode should encode the CU. Thus, when the correlated block in texture view is encoded with Skip mode ($M_{text}$ = Skip), and the RD-cost obtained by Skip and Intra-picture skip has a low value, there is a high probability of selecting Skip as the best encoding mode. Hence, it can be used to perform an early termination decision according to a threshold criterion, avoiding the evaluations of the remaining encoding modes.

### 5.2.1   Design of the Early Termination Block-Level Decision

Focusing on avoiding the excessive evaluations of the encoding modes in the traditional 3D-HTM encoding flow, the early termination block-level decision scheme privileges the modes with a high probability of being selected, with low computational effort and using fewer bits to encode blocks. The flowchart of the proposed scheme is presented in Fig. 5.5.

Based on our previous analysis, the scheme starts computing the RD-cost for Skip and Intra-picture skip modes for the given depth map CU. If the $M_{text}$ = Skip, the minimum RD-cost between Skip and Intra-picture skip modes is evaluated. If the minimum RD-cost divided by CU width size (Wsize) is lower than the threshold (TH) defined by in an offline analysis, then the previous evaluation mode that obtained the lowest RD-cost is selected. Otherwise, further mode evaluations are

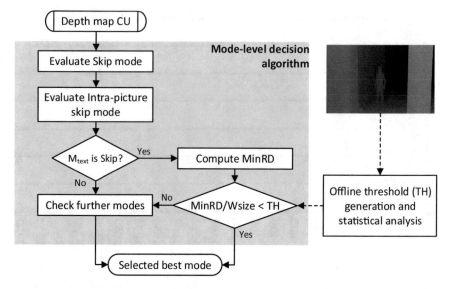

**Fig. 5.5** Dataflow model for early termination scheme

required, and the encoding process follows without simplification, requiring evaluating all encoding possibilities seeking for better encoding results. Equation (5.3) summarizes the decision process.

$$Decision = \begin{cases} lowest\ RD\,(DIS,\ SKIP), & \dfrac{MinRD}{Wsize} < TH \\ lowest\ RD\,(DIS,\ SKIP,\ INTER,\ MERGE,\ INTRA), & \dfrac{MinRD}{Wsize} \geq TH \end{cases} \tag{5.3}$$

In the best case, only Skip and Intra-picture skip are evaluated in our scheme. However, in the worst case, the RD-cost calculated by our solution for that depth map CU is the same as the conventional 3D-HTM encoding flow, without increasing the computational effort than the default approach would require.

The threshold values lead to light or aggressive solutions regarding both encoding time and video quality. Therefore, we employed an experimental analysis that evaluates seven scenarios to explain the impact of the threshold variation.

We selected the THs as evaluation targets, ranging from 100 to 400, with step 50. The corner values were selected by analyzing the PDF presented in Fig. 5.4, while step 50 was empirically selected because smaller steps would lead to a minimal variation. We evaluated 10 frames of the Undo_Dancer and GT_Fly sequences, which were selected randomly among the available sequences. Only two sequences were used in this analysis to avoid overfitting the designed scheme.

The threshold evaluations are displayed in Fig. 5.6, highlighting the percentage of early termination of encoding mode evaluation according to the BD-rate metric.

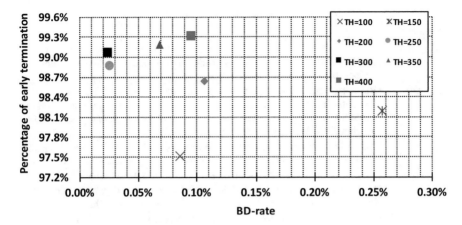

**Fig. 5.6** Percentage of early termination of the encoding mode evaluation according to the BD-rate impact

The percentage of early termination was computed as a division of the total cases our scheme skips the remaining evaluations by the total cases of $M_{text}$ = Skip.

Figure 5.6 shows the evaluated THs provide some operation points considering encoding efficiency and with a high percentage of early termination in the traditional 3D-HTM encoding flow (more than 96% of all evaluated cases). Based on this analysis, TH = 300 was selected as the best operating point for this early termination scheme.

## 5.3 Decision Trees for P- and B-Frames

The decision trees of the P- and B-frames definition process are similar to that previously used on the I-frames definition process. The features and attributes that could lead to effective I-frames splitting decisions were also evaluated for P- and B-frames. However, we analyzed new parameters since there are many different features in P- and B-frames when compared to I-frames, such as inter-frame or inter-view dependencies. These new parameters are listed in Table 5.1, along with their description.

Figure 5.7 shows the probability density functions regarding three of the selected parameters, to show the correlation among the parameters and the CU splitting decision. Figure 5.7a displays the probability of the CU not be split according to the Maximum mean Absolute Deviation of its 4 × 4 sub-blocks (MAD_4); Figure 5.7b shows the probability of the CU not be split according to the Neigh_depth, and Fig. 5.7c is the probability of the CU be split according to RelRatio. In the first two attributes (MAD_4 and Neigh_depth), low values tend to not split the encoding CU. On the other hand, low values of the RelRatio tends to split the encoding CU. Therefore, the knowledge of these attributes is crucial for achieving efficient decisions on current CU splitting.

**Table 5.1** Parameters evaluated in P- and B-frames and their description

| Parameter | Description |
| --- | --- |
| RD_MSM | RD-cost of merge/skip mode. It indicates the efficiency of using merge/skip mode; generally, high efficiency is obtained in slow motion or homogeneous regions. In this case, CUs tends to not be split |
| RD_DIS | RD-cost obtained in intra-picture skip mode; note the name of this mode is also found in the literature as DIS. Using it, the decision tree can verifies if the current encoding CU using intra-picture skip can be an attractive alternative to indicate if the current CU is composed of homogeneous regions |
| Ratio | RD-cost of the intra-frame prediction divided by the RD_DIS. It allows comparing the efficiency of the whole inter-frame prediction and DIS encoding mode |
| RatioInter | RD-cost of inter 2 N × 2 N evaluation divided by RD_MSM, which can be used to identify if the encoding region is complex or contains a high movement intensity. In this case, splitting the CUs tends to be interesting [8] |
| RelRatio | The normalized difference between the RD-cost of inter 2 N × 2 N and RD_MSM. This attribute is also used to identify if the region is complex or contains a high movement intensity |
| Neigh_depth | The average quadtree depth of the top, left, top-left, top-right, and co-located CTUs of both reference lists. This parameter is used based on the correlation of neighbor CUs |
| SKIP_flag | The binary flag notifies if the CU has been encoded using the skip mode. In cases the CU is encoded with skip, splitting the CU can be avoided |
| DIS_flag | The binary flag notifies if the CU is encoded using the DIS mode. In cases the CU is encoded with intra-picture skip, splitting the CU can be avoided |
| MAD | Maximum mean absolute deviation of smaller blocks inside of the current CU (4 × 4 up to current CU size) – Allows identifying the dispersion of the values inside a given CU |

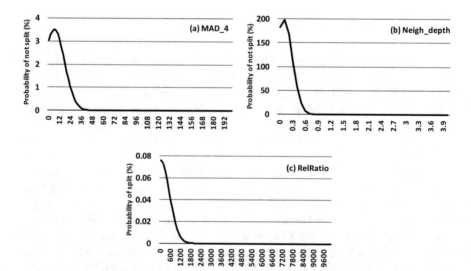

**Fig. 5.7** Probability density function of the current 64 × 64 CU in RA configuration not be split regarding (**a**) MAD_4 and (**b**) Neigh_depth, and (**c**) be split regarding RelRatio

The decision trees were trained using the Kendo video sequence. However, similar results were obtained repeating the same process with more video sequences or using a different video sequence. Table 5.2 illustrates the parameters used in the three decision trees for P- and B-frames. Table 5.2 also presents the IG of the attributes and the accuracy obtained by the designed trees, where the trees for 16 × 16, 32 × 32 e 64 × 64 CUs achieved an accuracy of 82.13%, 83.08% e 91.65%, respectively.

Figure 5.8 exemplifies the decision tree built by Waikato Environment for Knowledge Analysis (WEKA) [9] for 64 × 64 CUs. The built 32 × 32 and 16 × 16 CUs decision trees (that are omitted in this book) required six decision levels.

**Table 5.2** Parameters used in P- and B-frames decision trees

| Attributes | 64 × 64 | | 32 × 32 | | 16 × 16 | |
|---|---|---|---|---|---|---|
| | Used | IG | Used | IG | Used | IG |
| QP-depth | | – | × | 0.032 | × | 0.022 |
| R-D cost | × | 0.194 | × | 0.223 | × | 0.216 |
| RD_MSM | × | 0.198 | × | 0.182 | × | 0.154 |
| RD_DIS | | – | × | 0.214 | | – |
| Ratio | × | 0.450 | × | 0.258 | × | 0.228 |
| RatioInter | | – | × | 0.194 | | – |
| RelRatio | × | 0.483 | × | 0.194 | × | 0.139 |
| Neigh_depth | × | 0.493 | | – | | – |
| SKIP_flag | × | 0.257 | × | 0.068 | × | 0.050 |
| DIS_flag | | – | × | 0.063 | | – |
| VAR | × | 0.269 | × | 0.209 | | – |
| VAR_16 | × | 0.274 | | – | | – |
| MAD_4 | × | 0.271 | | – | | – |
| MaxDiff | × | 0.278 | | – | | – |
| Accuracy | 91.65% | | 83.08% | | 82.13% | |

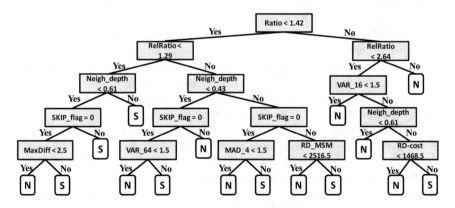

**Fig. 5.8** Decision tree for splitting decision in 64 × 64 CUs for P- and B-frames

## 5.4    Results and Comparisons

The three designed algorithms have been inserted into 3D-HTM and evaluated under CTC in RA configuration. Table 5.3 illustrates the results obtained with these algorithms. The performance difference in different videos execution occurs because each video has different characteristics as highlighted in Sect. 2.5.

The E-ADME algorithm reduces 3.2% the encoding time considering texture and depth map execution time with an increase of 0.1481% in the BD-rate; although it is not a significant reduction, if only ME/DE encoding effort is considered the 3D-HTM encoding time is reduced significantly as shown in Fig. 5.9, the SAD computation is reduced 68.2%, on average.

Comparing E-ADME to the other proposed techniques presented in Table 5.3, its results may seem worse than the others solutions; however, the BD-rate increase is still small and also, when designing real-time hardware it may be interesting to limit the ME/DE encoding effort by introducing this kind of technique. This reduction in SAD computations reduces the necessary accesses to the memory with reference samples and then, contributing to reduce the power dissipation. Consequently, the E-ADME scheme helps to the development of real-time encoding systems with negligible impact on the encoding efficiency.

Furthermore, one of the most significant bottlenecks in the ME algorithms relies on the fact that it has a high I/O communication with the reference memory. This I/O communication is strictly related to fetching information from previous reference frames to use in its SAD computation. Thus, reducing SAD computation in 68.2% reduce the memory access in the same proportion. Since the encoder bottleneck is the I/O communication with the memory [10], the proposed scheme tends to increase the performance of the entire 3D encoder.

Our block-level early termination decision algorithm reduces 13.6% the encoding effort considering the texture and depth execution time, increasing 0.1976% the BD-rate, with some variations around it. These variations happen due to the differences among the encoding sequences, and higher resolution video sequences tend to select more Skip than other modes in the texture coding. Figure 5.10 shows the significant evaluation skips obtained in our scheme of the cases when $M_{text}$ = Skip.

This result is better than the one achieved with E-ADME when considering only this tradeoff; however, its implementation in a real-time system is harder since it contains dependencies among encoding modules, limiting parallelism strategies. Besides, this scheme reduces 27.7% of the depth map encoding time. These results are explained by showing when this technique skips the remaining encoding modules (more than 96% of evaluations are skipped, on average).

Our last algorithm using a quadtree limitation based on machine learning reduced 52.0% the encoding time considering texture and depth map encoding time, with a drawback of increasing only 0.1588% the BD-rate in synthesized views. This was our best result in RA configuration when considering only the tradeoff between encoding timesaving and BD-rate. One drawback of this technique is the impossibility of encoding multiple block sizes together since the decision tree needs

**Table 5.3** Evaluations in the RA scenario

| Resolution | Video | E-ADME | | | Early termination Block-level decision | | Decision tree with machine learning | |
| --- | --- | --- | --- | --- | --- | --- | --- | --- |
| | | Synthesis only BD-rate | Timesaving | | Synthesis only BD-rate | Timesaving | Synthesis only BD-rate | Timesaving |
| 1024 × 768 | Balloons | 0.0206% | 3.6% | | 0.2084% | 12.3% | 0.0582% | 47.7% |
| | Kendo | 0.0246% | 3.4% | | 0.2948% | 11.1% | 0.0267% | 47.1% |
| | Newspaper_cc | 0.1057% | 3.7% | | 0.2620% | 11.8% | 0.0531% | 48.1% |
| | Average | 0.0503% | 3.6% | | 0.2551% | 11.7% | 0.0460% | 47.6% |
| 1920 × 1088 | GT_Fly | 0.0948% | 3.3% | | 0.0265% | 14.8% | 0.0747% | 52.2% |
| | Poznan_Hall2 | 0.2900% | 3.2% | | 0.3863% | 15.2% | 0.3921% | 59.4% |
| | Poznan_Street | 0.0713% | 3.6% | | 0.1452% | 15.8% | 0.1017% | 57.0% |
| | Undo_Dancer | 0.2036% | 2.3% | | 0.1327% | 14.0% | 0.5534% | 53.3% |
| | Shark | 0.3739% | 2.9% | | 0.1252% | 14.2% | 0.0103% | 51.4% |
| | Average | 0.2067% | 3.1% | | 0.1632% | 14.8% | 0.2264% | 54.7% |
| Average | | 0.1481% | 3.2% | | 0.1976% | 13.6% | 0.1588% | 52.0% |

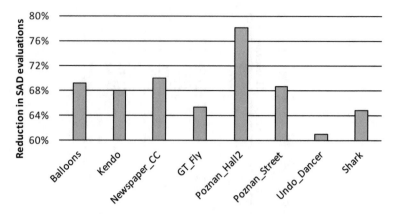

**Fig. 5.9** SAD computation reduction obtained using E-ADME

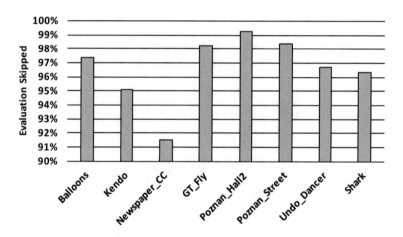

**Fig. 5.10** Evaluation skipped in block-level early termination scheme

a given level to be encoded before encoding or skipping a lower quadtree level. Figure 5.11 displays the percentage of CUs were not split according to the CU size and QP-depth value. The higher is the QP-depth value, the higher is the use of larger CUs and, consequently, the higher is the CUs with not split decisions. The best not splitting percentage reaches 97.7% with QP = 45 when processing 64 × 64 CUs, explaining the significant encoding time reduction. When only depth map coding is considered, this solution reduces 68.0% the encoding effort.

Table 5.4 compares our results to the related work. The E-ADME algorithm can be fairly compared only with [11], which is focused on reducing the DE encoding effort. In [11], timesaving results are not presented, but only the reductions in SAD computation. Besides, it is only focused on DE, while our work is focused on ME and DE. It obtained a BD-rate increase ranging between 0.3123% and 0.4803%, with a reduction between 32.7% and 61.8% in SAD computation. E-ADME provides lower BD-rate and requires fewer SAD computation (68.2%) compared to [11].

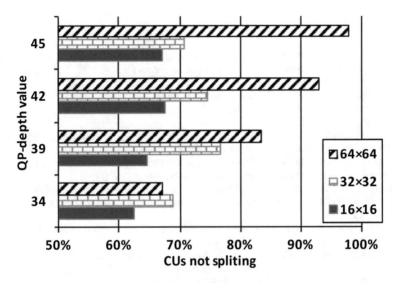

**Fig. 5.11**  Percentage of not splitting CUs using the designed quadtree limitation

**Table 5.4**  Comparison with related works in RA scenario

| Work | Decision level and focused tools | Synthesis only BD-rate | Encoding timesaving |
|---|---|---|---|
| E-ADME | Mode | 0.1481% | 3.2% |
| Early termination block-level decision | Block | 0.1976% | 13.6% |
| Quadtree limitation using machine learning | Quadtree | 0.1588% | 52.0% |
| V. Afonso et al. [11] | Mode | 0.3123% – 0.4803% | 32.7% – 61.8% (reduction in SAD computation) |
| R. Conceição et al. [12] | Block | 0.062% ~0.409% | 9.6% ~ 13.8% |
| J. Lei et al. [13] | Quadtree | 1.0240% | 52.0% |
| E. G. Mora et al. [14] | Quadtree | 2.0600% | 41.5% |

Our early termination block-level decision can be compared to [12] that also proposes a block-level decision. Its results reduce the encoding time between 9.6% and 13.8% with a BD-rate increase of 0.062%–0.409%. This high encoding effort reduction is similar to the one provided by our algorithm; however, their solution requires more than the double of the BD-rate increase, showing the advantage of our solution.

Comparing our quadtree limitation using machine learning with the related work focusing on quadtree-level, our solution achieves competitive encoding timesaving, reaching the same results than [13] and better results than [14]. Additionally, our solution reached a much smaller BD-rate increase than those solutions.

# References

1. Sanchez, G., M. Saldanha, B. Zatt, M. Porto, L. Agostini, and C. Marcon. 2017. Edge-aware depth motion estimation—A complexity reduction scheme for 3D-HEVC. In: European Signal Processing Conference, 1524–1528.
2. Saldanha, M., G. Sanchez, C. Marcon, and L. Agostini. 2018. Block-level fast coding scheme for depth maps in three-dimensional high efficiency video coding. *Journal of Electronic Imaging* 27 (1): 010502.
3. ———. 2019. Fast 3D-HEVC depth maps encoding using machine learning. *IEEE Transactions on Circuits and Systems for Video Technology* 1–1: 12.
4. Zhu, S., and K.-K. Ma. 2000. A new diamond search algorithm for fast block-matching motion estimation. *IEEE Transactions on Image Processing* 9–2: 287–290.
5. Saldanha, M., G. Sanchez, B. Zatt, M. Porto, and L. Agostini. 2015. Complexity reduction for the 3D-HEVC depth maps coding. *IEEE International Symposium on Circuits and Systems* 2015: 621–624.
6. ———. 2017. Energy-aware scheme for the 3D-HEVC depth maps prediction. *Journal of Real-Time Image Processing* 13 (1): 55–69.
7. Sanchez, G., M. Saldanha, G. Balota, B. Zatt, M. Porto, and L. Agostini. 2014. Complexity reduction for 3D-HEVC depth maps intra-frame prediction using simplified edge detector algorithm. In: IEEE International Conference on Image Processing, 3209–3213.
8. Correa, G., P.A. Assuncao, L.V. Agostini, and L.A. da Silva Cruz. 2015. Fast HEVC encoding decisions using data mining. *IEEE Transactions on Circuits and Systems for Video Technology* 25 (4): 660–673.
9. Hall, M., E. Frank, G. Holmes, B. Pfahringer, P. Reutemann, and I.H. Witten. 2009. The WEKA data mining software: An update. *ACM SIGKDD Explorations Newsletter* 11 (1): 10–18.
10. Tuan, J.-C., T.-S. Chang, and C.-W. Jen. 2002. On the data reuse and memory bandwidth analysis for full-search block-matching VLSI architecture. *IEEE Transactions on Circuits and Systems for Video Technology* 12 (1): 61–72.
11. Afonso, V., A. Susin, M. Perleberg, R. Conceição, G. Corrêa, L. Agostini, B. Zatt, and M. Porto. 2018. Hardware-friendly unidirectional disparity-search algorithm for 3d-hevc. *IEEE International Symposium on Circuits and Systems*: 1–5.
12. Conceição, R., G. Avila, G. Corrêa, M. Porto, B. Zatt, and L. Agostini. 2016. Complexity reduction for 3D-HEVC depth map coding based on early skip and early DIS scheme. *IEEE International Conference on Image Processing* 2016: 1116–1120.
13. Lei, J., J. Duan, F. Wu, N. Ling, and C. Hou. 2018. Fast mode decision based on grayscale similarity and inter-view correlation for depth map coding in 3D-HEVC. *IEEE Transactions on Circuits and Systems for Video Technology* 28 (3): 706–718.
14. Mora, E.G., J. Jung, M. Cagnazzo, and B. Pesquet-Popescu. 2014. Initialization, limitation, and predictive coding of the depth and texture quadtree in 3D-HEVC. *IEEE Transactions on Circuits and Systems for Video Technology* 24 (9): 1554–1565.

# Chapter 6
# Conclusions and Open Research Possibilities

This book proposed several algorithms at different granularity levels to provide timesaving at intra- and inter-frame prediction of depth maps. This book was motivated by the requirements of depth maps together with the texture views in achieving 3D videos with high quality and high compression rates. The 3D-HEVC standardization proposed and adopted several new tools to encode depth maps efficiently, increasing the effort spent to encode 3D videos. Consequently, new challenges arise regarding designing efficient and efficacious algorithms compatible with 3D-HEVC, especially aiming real-time system and low-power design.

This book contains timesaving algorithms with minor impact in the encoding quality to the most time-consuming encoding tools, contributing to the designing of efficient 3D video coding systems. The main contributions of this book are the encoding effort reduction on the intra-frame, inter-frame, and inter-view predictions.

Chapter 4 englobes a deep encoding time and mode distribution analysis, which inspired the design of four new algorithms for encoding effort reduction at intra-frame prediction: Depth Modeling Mode-1 Fast Pattern Selection (DFPS), Pattern-based Gradient Mode One Filter (P&GMOF), Enhanced Depth Rough Mode Decision (ED-RMD), and quadtree limitation with machine learning. The first two algorithms, DFPS and P&GMOF, were designed focusing on reducing the Intra_ Wedge encoding effort. ED-RMD was planned for reducing the encoding effort spent in Transform-Quantization (TQ) and Direct Component-only (DC-only) evaluations. The quadtree limitation with machine learning was designed as a high-level decision approach, where the encoding effort spent in lower levels of the quadtree were skipped frequently when the encoding scenario tends to require larger blocks. Experimental results demonstrate that the proposed algorithms reduce 6.5–52.4% the encoding effort, affecting only 0.0119–0.1770% the Bjøntegaard delta-rate (BD-rate) of the synthesized views. These results are very competitive when compared to related work.

Chapter 5 presents the contributions related to the encoding effort reduction for inter-frame prediction. It shows the design of three algorithms with different levels

© Springer Nature Switzerland AG 2020
G. Sanchez et al., *Algorithms for Efficient and Fast 3D-HEVC Depth Map Encoding*, https://doi.org/10.1007/978-3-030-25927-3_6

of granularity. The depth map simplicity inspired the design of Edge-Aware Depth Motion Estimation (E-ADME), which allows speeding up the Motion Estimation (ME) of the depth maps when encoding homogeneous regions while keeping the initial evaluation when encoding edges. This algorithm is based in an edge classifier that speeds up the evaluation of the homogeneous region in intra-frame prediction. The second algorithm design a block-level scheme for early termination decisions. Based on the results of the most used modes (i.e., Skip and Intra-picture skip), this algorithm determines if the remaining evaluations inside an encoding block are required or not. Therefore, when Skip and Intra-picture skip results are satisfactory, the block is encoded without the remaining evaluations, and the encoding process is finalized. The last algorithm employs specific decision trees using machine learning for pruning the quadtree at inter-frame prediction. This work is similar to the intra-frame prediction, however, in inter-frame prediction, more attributes were evaluated to achieve better results; i.e., the experimental results show an encoding effort reduction between 3.2% and 52.0%, with a drawback of increasing the BD-rate in a range from 0.1481% to 0.1976% in the synthesized views.

In summary, this book shows the design of several algorithms and enhancements in the encoder system. The experimental evaluations demonstrated significant time-saving with minor impact in the quality of synthesized views can be achieved in encoding depth maps when considering their characteristics. Besides, several points can still be explored for achieving higher performance in coding depth maps as described next Section.

## 6.1  Future Works and Open Research Possibilities

Several research possibilities were not covered by the literature yet in the coding of depth maps. For instance, the highest timesaving algorithms in both intra- and inter-frame prediction obtained around 50% of encoding effort reduction, which is already a significant reduction; however, for allowing real-time applications using 3D-HEVC encoders, it is necessary a higher reduction in encoding effort or obtaining significant speed up by using parallelism in the 3D-HEVC encoder. Therefore, there still significant space for outcoming works to allow this real-time design.

Higher reductions can be achieved on the algorithmic exploration by using more sophisticated machine learning techniques such as deep learning. For allowing a high performance, it is necessary the design of new hardware architectures for the 3D-HEVC module; however, there are only a few works in the literature designing small solutions for the 3D-HEVC coding of depth maps. Consequently, there is still room for design fast and low-power systems exploring the parallelism in the depth map encoding. Besides, there is also room for speeding up the 3D-HEVC using architectures with massive parallelism such as Graphic Processor Units (GPU) or Multiprocessor System-on-Chip (MPSoC), which were not searched in depth. Besides, the usage of the tiles technique for parallelism computing was not much explored in both texture and depth map encoding.

# Index

© Springer Nature Switzerland AG 2020
G. Sanchez et al., *Algorithms for Efficient and Fast 3D-HEVC Depth Map
Encoding*, https://doi.org/10.1007/978-3-030-25927-3

Printed in the United States
By Bookmasters